ETHICS AND THE WILL

VIENNA CIRCLE COLLECTION

HENK L. MULDER, *University of Amsterdam, Amsterdam,
The Netherlands*

ROBERT S. COHEN, *Boston University, Boston, Mass., U.S.A.*

BRIAN MCGUINNESS, *University of Siena, Siena, Italy*

RUDOLF HALLER, *Charles Francis University, Graz, Austria*

Editorial Advisory Board

ALBERT E. BLUMBERG, *Rutgers University, New Brunswick, N.J., U.S.A.*

ERWIN N. HIEBERT, *Harvard University, Cambridge, Mass., U.S.A*

JAAKKO HINTIKKA, *Boston University, Boston, Mass., U.S.A.*

A. J. KOX, *University of Amsterdam, Amsterdam, The Netherlands*

GABRIËL NUCHELMANS, *University of Leyden, Leyden,
The Netherlands*

ANTHONY M. QUINTON, *All Souls College, Oxford, England*

J. F. STAAL, *University of California, Berkeley, Calif., U.S.A.*

FRIEDRICH STADLER, *Institute for Science and Art, Vienna, Austria*

VOLUME 21

VOLUME EDITOR: BRIAN McGUINNESS

FRIEDRICH WAISMANN
JOSEF SCHÄCHTER
MORITZ SCHLICK

ETHICS AND THE WILL

Essays

Edited and with an Introduction by
BRIAN McGUINNESS
and
JOACHIM SCHULTE

Translations by
HANS KAAL

KLUWER ACADEMIC PUBLISHERS
DORDRECHT / BOSTON / LONDON

Library of Congress Cataloging-in-Publication Data

```
Ethics and the will : essays / by Friedrich Waismann, Josef Schächter,
  Moritz Schlick ; translations by Hans Kaal ; edited and with an
  introduction by Brian McGuinness and Joachim Schulte.
       p.   cm. -- (Vienna circle collection ; v. 21)
    Includes bibliographical references and index.
    Contents: Introduction -- The main ideas of the theory of values /
  Moritz Schlick -- The meaning of pessimistic sentences / Josef
  Schächter -- Notes on problems of ethics and philosophy of culture /
  Josef Schächter -- Ethics and science / Friedrich Waismann -- Will
  and motive / Friedrich Waismann.
    ISBN 0-7923-2674-1 (alk. paper)
    1. Ethics.  2. Free will and determinism.  3. Vienna circle.
  I. Waismann, Friedrich.  II. Shechter, Y., 1901-    .  III. Schlick,
  Moritz, 1882-1936.  IV. McGuinness, Brian.  V. Schulte, Joachim.
  VI. Series.
  BJ1114.E76  1994
  170--dc20                                                    93-47902
```

ISBN 0-7923-2674-1

Published by Kluwer Academic Publishers,
P.O. Box 17, 3300 AA Dordrecht, The Netherlands.

Kluwer Academic Publishers incorporates
the publishing programmes of
D. Reidel, Martinus Nijhoff, Dr W. Junk and MTP Press.

Sold and distributed in the U.S.A. and Canada
by Kluwer Academic Publishers,
101 Philip Drive, Norwell, MA 02061, U.S.A.

In all other countries, sold and distributed
by Kluwer Academic Publishers Group,
P.O. Box 322, 3300 AH Dordrecht, The Netherlands.

printed on acid-free paper

All Rights Reserved
© 1994 Kluwer Academic Publishers
No part of the material protected by this copyright notice may be reproduced or
utilized in any form or by any means, electronic or mechanical,
including photocopying, recording or by any information storage and
retrieval system, without written permission from
the copyright owner.

Printed in The Netherlands

TABLE OF CONTENTS

INTRODUCTION	vii
MORITZ SCHLICK / The Main Ideas of the Theory of Values	1
JOSEF SCHÄCHTER / The Meaning of Pessimistic Sentences	7
JOSEF SCHÄCHTER / Notes on Problems of Ethics and the Philosophy of Culture	19
FRIEDRICH WAISMANN / Ethics and Science	33
FRIEDRICH WAISMANN / Will and Motive	53
INDEX OF NAMES	139

INTRODUCTION

The present volume unites contributions by the leading figure of the Vienna Circle and by two of his closest associates, contributions that deal with an area of thought represented, indeed, in this Collection but certainly not the central one in the common picture of the Circle's activities. It is no accident that an interest in ethics and the philosophy of action was particularly marked in what Neurath was apt to call the right wing of the Circle. For them, as for Wittgenstein (the respected mentor of Schlick and Waismann in particular), the advance to be hoped for in philosophy consisted not solely in freeing natural science from a confused sense of dependence on speculative metaphysics but also in seeing that other areas of language and action had to be thought about in their own terms, which were neither those of natural science nor those of philosophy as traditionally conceived. The scepticism of Schlick about the programme of Unified Science was well known: *Einheizwissenschaft* he called it, as it might be 'boozified science'. And in sober truth the programme sometimes masked a left-wing set of values taken (surely illogically) for granted, though the members of the Circle entertained a wide range of political views.

Schlick's own contribution to the present volume is a section from the notes for one of his final lecture series, for sight of which we warmly thank the only surviving contributor to our volume, Dr Joseph Schächter: Schlick's grandson Dr G.M.H. van de Velde has kindly consented to their publication. This section poses the problem we have outlined: there are questions and a need for clarification in ethics, but these no more demand a metaphysical solution than does a similar situation in epistemology. Here, as in his earlier *Problems of Ethics*,[1] Schlick sets his face against the whole process, most obvious in Kant, of making the concept of value absolute. One might say that for Schlick there is no unhypothetical imperative. He seems, as in his previous volume, to think that the determination of the conditions which legitimize a moral imperative is essentially a task for psychology. In passing Schlick also adumbrates the criticism one might expect, that the problem of freedom of the will results from ill-formulated questions. These problems or pseudo-problems of the justification of ethics and of the correct understanding of the phenomenon, or rather the concept, of willing are confronted, though not quite in the way Schlick indicates, in the contributions by his two pupils. Though they otherwise diverge from his book and (as far as we can tell) these lectures, they share, of course, his view that the problems of epistemology do not exhaust philosophy.

First we follow these few pages with two contributions that Dr Schächter has kindly made available from his own work. He is no stranger to readers of our collection, since his *Prolegomena to a Critical Grammar* constitute our second volume.[2] A relatively full bibliography of his works will be found there, which gives some idea of how after an apprenticeship, so to speak, as an assistant to Schlick, he went in difficult years to Israel and carried further the work of a philosopher, with, eventually, considerable effect on the intellectual and educational life of the State of Israel. It is of interest in our present context to note that a major theme of his later writings was the possibility within the Jewish heritage of a 'philosophy of life'. The nature and the possibility of such a philosophy and its independence (as he maintains) of myth and religion is made clear in the two essays printed here. (One of these, 'The Meaning of Pessimistic Sentences' originally appeared, in German, in the 1938 volume of *Synthese*, now a Kluwer review. The other, written in 1937, has not previously been published.)

These essays exemplify an approach to ethics in which it is no longer viewed as a matter of psychological fact (as in Schlick's own writings) but is seen instead as one of the possible, indeed natural, strata of language-use. In this way the place and the interrelations of myth, of religion, and of culture are derived from a philosophy of language of considerable sophistication, especially for its date. Another bold step at this point is Schächter's illumination of ethics by logic and vice versa. The parallels drawn would have deserved more timely publication but at any rate are now offered for exploration. The topic is reminiscent of Wittgenstein's remark, also published long after it was made, that ethics must be a condition of the world, like logic.[3]

It is an interesting commentary on the Circle as a whole that so many of its members, in desperate times and in a place where despair seemed concentrated, retained in their writings on human conduct a faith – one is tempted to say a blind faith – in the powers of clear thinking. Obviously this was not true of Neurath with his blend of scientism and Marxism, but Schlick himself in his essay on the sense of life[4] and Menger in his volume on choice[5] are examples. Schächter too concludes with a portrait full of hope. The heir to the religious man of the past will be 'the essential man', who will in no circumstances inwardly accept tyranny. Such a man will, of course, as Schächter points out, be Socratic, but readers will also be reminded of the Lutheran-turned-Catholic mystic, Angelus Silesius and his injunction, *'Mensch, werde wesentlich!'*.

The remaining two contributions, and the bulk of the volume, are two works of Friedrich Waismann, whose *Philosophical Papers* was the eighth volume of our Collection.[6] The identification and publication of these further works was the result of a closer examination of Waismann's *Nachlass* on the occasion of its cataloguing by one of the present editors (J.S.). He it was who prepared them for publication in German, under the title *Wille und Motiv*, Reclam, Stuttgart, 1983, and the remarks which follow are a translation, with slight adaptations, of part of his 'Epilogue' to that publication. For the re-use of this material and indeed for the opportunity to bring Waismann's work to the English-reading public, we are much indebted to the house of Reclam as well as to Waismann's literary heirs, Sir Isaiah Berlin, O.M., and Sir Stuart Hampshire. (We do not offer here an account of Waismann's life, for which purpose Anthony Quinton's introduction to our former volume will admirably serve.[7])

Waismann's work can be divided naturally into three periods, though it must not be forgotten that his earlier ideas may also recur in his latest work. The first period is one of close collaboration with Wittgenstein and the first stages of his work on *Logik, Sprache, Philosophie*.[8] During this period Waismann published his first articles in *Erkenntnis*, the journal of the Vienna Circle. The second period runs from his estrangement from Wittgenstein to his move to Oxford. During this period Waismann completed the German version of *Logik, Sprache, Philosophie* as we know it today [later published in English under the title *The Principles of Linguistic Philosophy*]. At the same time he 'plundered' his voluminous manuscript to publish separate parts of it as articles, first in *Erkenntnis* and then in *Synthese*. Waismann's *Einführung in das mathematische Denken*, the only book he published in his lifetime, appeared in 1936 and has been translated in the meantime into several languages.[9] The third period is marked by Waismann's teaching activities at Oxford and his publication of such significant papers as 'Verifiability', 'Language Strata', 'Analytic-Synthetic' and 'How I See Philosophy'.[10]

The essay 'Ethics and Science', here published for the first time, was written towards the end of his second period. In it Waismann discusses with exemplary clarity problems about the justification of moral precepts and the foundations of ethical systems, and raises questions about the place of ethics in an age of science. He argues forcefully against intuitionist attempts to found ethics and criticizes the principle of evidence (in its non-ethical as well as ethical applications). He compares opposing views on happiness and applies his knowledge of the logic of language to the concepts of truth and moral obligation. The conclusion Waismann draws from these discussions is that it is impossible to give a moral code a rational foundation: 'Morality, like religion, is something one can only *profess*', a conclusion clearly reminiscent of Wittgenstein's view, as expressed in a conversation with Waismann in 1930: 'At the end of my lecture on ethics I spoke in the first person: I think that this is something very essential. Here there is nothing to be stated any more; all I can do is to step forth as an individual and speak in the first person.'[11]

However, in Waismann's manuscript on the will, Wittgenstein's influence is less obvious. This treatise dates presumably from the mid-1940s and hence from the beginning of Waismann's third period, and the content indicates that Waismann owes some of his insights to his discussions with Ryle. This manuscript, called conventionally 'Freedom of the Will',[12] is a finished draft, but deals with freedom of the will only in the first few pages, where Waismann rejects the classical problem of whether our actions are determined or undetermined as an improper question. It could perhaps be said that he rejects it as a 'pseudo-problem' even though he himself does not use that expression, so typical of the Vienna Circle. The text deals basically with the concepts 'will' and 'motive', and in the final chapter Waismann actually speaks of his 'treatise on the will' (p. 132). If we speak here of the *concepts* 'will' and 'motive', it is because the main goal of Waismann's undertaking is conceptual clarification or conceptual analysis. To our mind, his results can best be characterized as the beginnings of a theory of action – 'beginnings' because the theory is not fully developed, and 'theory of *action*' because Waismann always insists that the concepts to be analysed be examined to see whether they can be applied in describing and explaining an action as a whole.

Why does Waismann reject the traditional problem of the freedom of the will as – in effect – a pseudo-problem? True to his intellectual origins in the Vienna Circle, he takes our criterion for an event's being determined to be whether it is predictable on the basis of (scientific) laws. Especially if we know a person well, we can indeed predict his actions within certain limits, and we may perhaps do so on the basis of lawlike statements, but on Waismann's view the notion of predictability does not do justice to human action or to our concept of it. We have a picture of the way an action is initiated and performed which makes it implausible and even senseless to apply the scientific concept of something's being determined, and Waismann accepts this picture as correct, at least in rough outline; it shows 'a person at the moment of decision ... before him is an open cone of possibilities, some more probable, others less so, but all of it uncertain and fluid, without sharp boundaries' (p. 57). Since Waismann accepts this picture as correct, he has to drop the criterion of predictability; and since it is on his view the *only* criterion for something's being determined, we have to give up the whole classical problem because we are no longer able to distinguish between those human actions that are determined and those that are not.

However, this rejection of the criterion of being determined does not mean that we cannot say anything about the will, nor that we are forced to rely on subjective experience, i.e. introspection. On the contrary; while admitting that such an inner experience, especially as remembered after the fact, has to be taken into account in explaining intentional, willed action, Waismann repeatedly points out the dangers of being misled by an appeal to introspection. On the one hand, he denies that introspection is relevant to deciding whether something was willed, and on the other he points out that in order to find out whether I willed or not I have to treat myself the way another person would: I have to wait and see how I will act in future, what I am able to do and how I will behave, in order to know whether the step I took at the time was really willed.

One of the main goals of Waismann's investigation is to answer the question when we are justified in speaking of willing as opposed to, say, wishing. This question seems to ask for necessary and/or sufficient conditions, but on Waismann's view this is impossible, and for more than one reason. He first proposes to distinguish willing from wishing by saying that while wishing may remain purely subjective, or a mere idea, willing has to manifest itself: i.e. whether something was really willed is shown only in action. Waismann sums up this thesis in the slogan: 'the will is the act' (p. 58). However, in the course of the investigation it soon appears that this slogan gives neither necessary nor sufficient characteristics of willing. In short, action is not necessary because we are also justified in speaking of willing when we do not succeed in performing an action, i.e. when there is no demonstrable event that could be called the manifestation of the will; and action is not sufficient because we can speak of actions (e.g. in the case of routine or automatic actions) when there is no question of willing or voluntary action. The slogan 'the will is the act' is thus at bottom no more than a highly fallible criterion for distinguishing willing from wishing.

Among the essential characteristic marks of willing, Waismann emphasizes especially making an effort, or overcoming a resistance: we can speak of willing only when there is at least a minimal resistance to be overcome, something that occupies our attention. On the other hand, the resistance must not be too great, for otherwise there could again be no willing. Thus while I can *wish* to play the Hammerklavier Sonata, even though up to now I have got only as far as the Sonata facile, I cannot *will* it: there is a glaring discrepancy between what I wish and what I am able to do.

However, Waismann is not just concerned with the question when we are justified in speaking of 'willing' or willed action. He also wants to tackle the problem of how one gets to do a willed action, i.e. what kind of answers are appropriate to the question 'Why did he do it?'. On Waismann's view, a satisfactory answer to such a why-question usually gives the *motive* of the action.

But what are motives? One thing is certain: here too introspection does not get us any further, for what I perceive in myself when I try to find out why I will or willed to do something may be so diverse, changeable and insignificant that it cannot be expected to be informative. And if upon reflection I give a proper answer to the question why, by saying, e.g., 'I did it for the money', then this supposedly correct answer does not correspond to an introspectible element in my consciousness. Inner processes and motives need not stand in any recognizable relationship, as Ryle also stressed in his classic treatise *The Concept of Mind*,[13] and this is especially clear in cases in which the motive arises from some disposition or as it is called in ordinary language, character trait. Waismann's example is: 'He acted out of jealousy'. It is clear that an action done out of jealousy can be performed without any jealousy-related ideas, images or feelings. I.e., a motive like jealousy can remain *unconscious*.

Not so in cases in which I answer the question about the motives of my action by stating an aim or an intention to be fulfilled. Thus I may answer the question why about a certain action by saying: 'I kicked him in order to wake him up' or some such thing. In all such cases, the agent must be assumed to have an idea of the aim of his action or the object of his will before his mind. In other words, he must be *conscious*, however vaguely, of his motive.

Waismann's terminology makes use of this distinction between necessarily conscious motives on the one hand and possibly unconscious motives on the other: Waismann calls the former motives (i.e. those connected with intention and consciousness) 'purposes' [*Beweggründe*] and the latter (such as jealousy, ambition, etc.) 'drives' [*Triebfedern*]. But as Waismann shows by discussing some examples, this distinction does not amount to an exhaustive dichotomy that can be extended to all cases. Thus Waismann argues that there are motives which do not give rise to an aim or conscious intention, but are nevertheless conscious: e.g. 'acting in anger', 'in high spirits', etc. The motives designated by these concepts differ from 'grounds' mainly by the lack of a specific goal for the action; and they differ from 'drives': on the one hand, by being necessarily conscious, and on the other, by their transitory nature; anger, fury, etc. usually pass quickly, whereas jealousy, cupidity and the like tend to persist. This third class of motives, which differ clearly from 'drives' and 'purposes', Waismann calls 'impulses' [*Antriebe*].

We give motives to explain why an action was performed. This applies to cases where we want to explain or justify our own actions as well as cases where we want to get to understand another person's actions. We bring in the will to say that an action was performed, on the one hand, voluntarily, i.e. without external constraint, and on the other hand, not automatically or even unconsciously. Thus it looks as if our ordinary picture of an action was that of a series of events beginning with a motive chosen and adopted by the will, and ending with the action proper, which is produced by the will on the basis of its decision.

Waismann makes it clear that there are many things wrong with this picture. Let us visualize the direction of the supposed process:

$$\text{MOTIVE} \rightarrow \text{WILL} \rightarrow \text{ACTION}$$

Must we not ask ourselves what the will bases its decision on when it chooses a particular motive? And what can it be if not a further motive? If we ask such questions, we see at once that there are many starting-points for infinite regress arguments, as we know them, e.g., from Ryle's work. Waismann makes use of similar arguments, e.g. when he writes:

> If we now ask what determines the will, the answer seems to be; the motive. For we generally conceive of a motive as what moves us to do something, and hence, as what sets the will in motion. So the will chooses and determines the motive; but how can it do that if it is determined by the motive? Further, if the will is determined by some other thing, how can the *will* be what moves us? For then the *motive* would have the power to move, and the will would really be quite superfluous (p. 107f).

There is really no way to say it more clearly: there is something fishy about our picture in which the will chooses a motive and accordingly produces an action, for if we look at the matter as a temporally ordered causally determined process, we see that there is really no need for the will.

On the other hand, we have seen that we have only the beginning of an understanding of the nature of motives: we have no adequate definition of the term, and we are unable to give a complete list of kinds of motives. Moreover, as Waismann points out, motives *change*, i.e., today the motive of an action seems to be *this*, tomorrow *that*. Thus while I believe today that I acted from altruistic motives, I may see later that there was quite a bit of egotism involved. Or, seen from different perspectives, to me the motive of action may be *this* and to you *that*. Thus Joan may believe that I married her for her money, while I myself attribute it to my lack of a strong ego. Motives thus depend on the way we look at things. True, we cannot choose them arbitrarily, for sometimes we can justify the choice of a given motive and sometimes we cannot. But there is no final court of appeal which would definitely settle the question which motive was the true one. As Waismann puts it, motives are 'interpretations'. This is not to speak in favour of a rank pluralism of explanations, but simply to indicate the limits of our knowledge. Besides, not all motives are in the same situation. Waismann points out that in special cases we can collect countless data about how an action was initiated and performed, in such a way as to come at least close to a *causal* explanation of it. Waismann writes: 'In this case asking for the motive is much more like asking for the *cause*' (p. 130). But this is hardly the standard case in which we have to cope with the following situation: 'Since [...] motives are unstable and fade away on a critical view, it is better from the outset not to conceive of them as existing things, but to observe what really happens when we act and then judge our own action' (p. 129).

What conclusions can we draw from these considerations? On the one hand, Waismann says that neither will nor motive is a phenomenon that can be isolated and explained by disregarding the context of an action. They are not independent entities whose nature we can recognize by careful inspection, but essentially dependent elements of our action, which we can explain functionally, and whose existential presuppositions we can characterize at least in rough outline. We can specify under what condition we will speak of 'willing' or willed action, and we can also describe, analyse and classify kinds of motives. These possibilities should clear up some obscurities, but at the same time warn us to be modest in our knowledge claims.

NOTES

[1] New York, 1939 (German original: *Fragen der Ethik*, Vienna: Springer, 1930).
[2] Dordrecht/Boston: Reidel, 1973, with a Foreword by J.F. Staal.
[3] Notebook remark for 24 July 1916, first published in Ludwig Wittgenstein *Notebooks 1914–1916*, Oxford: Blackwell, 1961, p. 77.
[4] 'Vom Sinn des Lebens', pp. 331–54 in *Symposion* 1, 1927, (E.T. in Moritz Schlick, Philosophical Papers vol.II, Dordrecht/Boston: Reidel, 1979, pp. 112–129).
[5] *Morality, Decision and Social Organization*, Dordrecht/Boston: Reidel, 1974.
[6] Dordrecht/Boston: Reidel, 1977, with an Introduction by Anthony Quinton.
[7] See preceding note. Readers will also find instructive the introduction by Wolfgang Grassl to Friedrich Waismann, *Lectures on the Philosophy of Mathematics*, Amsterdam: Rodopi, 1982 (edited by Grassl).
[8] Friedrich Waismann, *Logik, Sprache, Philosophie*, with a preface by Moritz Schlick, edited by Gordon P. Baker and Brian McGuinness with the collaboration of Joachim Schulte, Stuttgart: Reclam, 1976.
[9] Friedrich Waismann, *Einführung in das mathematische Denken*, edited by Friedrich Kur, Munich: Deutscher Taschenbuch Verlag, 1970 (first published Vienna: Gerold, 1936; E.T. *Introduction to Mathematical Thinking*, New York: Ungar, 1951).
[10] See the bibliography in J.S.'s article mentioned in footnote 12.
[11] Friedrich Waismann, *Wittgenstein and the Vienna Circle*, edited by Brian McGuinness, Oxford: Blackwell, 1979, p. 117.
[12] J.S. also kept this title in his report on Waismann's Nachlass in: *Zeitschrift für philosophische Forschung* (1979), p. 108–140. The manuscript translated here is listed there under Catalogue No. L.2. The conventional title 'Freedom of the Will' is presumably due to the fact that Waismann kept the manuscript in a folder entitled *Willensfreiheit*.
[13] Gilbert Ryle, *The Concept of Mind*, London: Hutchinson, 1949.

MORITZ SCHLICK

THE MAIN IDEAS OF THE THEORY OF VALUES

MORITZ SCHLICK

LECTURES, CHAPTER 4:
THE MAIN IDEAS OF THE THEORY OF VALUES

From our explanations up to now we still do not know what philosophy is. But we have established two things: philosophy must yield not only knowledge but also wisdom. Let us now try to get an overall view of the basic ideas behind the search for wisdom, which has generally been subsumed under the term 'ethics' or 'moral philosophy' or 'practical philosophy'. The question about what is valuable must, logically speaking, be answered *before* man acts; it is therefore tantamount to the question; how shall man act? As we shall see, the answer may lead us also into metaphysics.

Such questions go back to Socrates, who was the first to deal with them: this is what was meant by saying (as Cicero did) that Socrates brought philosophy down to earth. The oldest philosophy dealt with the structure of the universe and has accordingly been called natural philosophy. Socrates on the other hand dealt with man, i.e. with questions concerning man; he asked for example: 'How should I act?' In answering this question, ancient philosophers were all basically of the same opinion. They said that the goal of any action is happiness (*eudaemonia*). For they reasoned roughly as follows. Any action has a purpose. An action completely without sense or purpose is inconceivable. Thus we climb a mountain in order to enjoy the view, write a letter to someone in order to get him to do something, etc. It is clear that if we continue with the question about the purpose of any such action and ask why man acts at all, we get the answer that *man acts in order to be happy*. It is, however, difficult to say what it means to 'be happy'. But we do not want to criticize this view; we are simply stating the fact that for ancient philosophers the valuable coincided with the good.

The word 'good' is used in various ways. All that Socrates does is try to combine all the meanings of the word 'good' and define the ethically good as a species of the good as such. We speak, e.g., of a good shoemaker, a good pair of boots, a good helmsman, a good ship, etc. and mean by it something that distinguishes them. All of them are as they should be, as one wishes them to be, and hence, such as to make me happy. A man is morally good if he acts in a certain way, namely virtuously, for on Socrates' view this kind of action leads to happiness. Socrates tries to combine the many meanings of the word 'good'; Plato's idea of the good is a continuation of this idea. Socrates reasons further: The ways that lead to happiness and hence constitute virtuous actions can be *learned*, for they are practical guidelines. As soon as we know what is good, we also do it, since it leads to happiness, which is what all men strive for. All immorality is thus based on ignorance. This view has been called 'intellectualism'.

We have seen that the question about the ultimate goal of action arises when one continues asking questions of everyday life beyond their area of application. Ancient philosophers agree in answering the question about the one and only ultimate, original, independently existing value by saying that the valuable as such is happiness. There is no question of any other ultimate value. To ancient philosophers the real problem is not this, but how to achieve happiness; to them this is what is meant by asking what the good is. To this question they give different answers, e.g. among others: Happiness consists in the soul's dedicating itself to rational activity (Aristotle).

The Stoics on the other hand answer this question by saying that happiness consists in living according to nature. This sentence is, however, in need of interpretation. What is meant by living according to nature? Does this mean adapting to nature or following one's own nature? The expression 'to follow one's own nature', which implies satisfying man's mental side, brings in for the first time what is called the 'seemly' or 'fitting'. *The pleasant good is being contrasted with the dutiful good.*

Epicurus answers the above question by saying that the good consists in imperturbability of mind (*ataraxia*), a calm enjoyment suited to the temperament of a sage. But virtue and justice are the *precondition* of this imperturbability of mind. This contrasts with Socrates' saying that virtue is already happiness itself. (In what follows we will not be criticizing either of these views.) The view that virtue and justice are only the precondition of a happy life is based on the empirical fact that a virtuous life need not always be happy. On the other hand, Epicurus takes the view that a virtuous life does at least *deserve* happiness. These ideas are also basic to Christianity, except that Christianity defers the reward for a virtuous life until the hereafter. The concept of value now takes on a metaphysical colouring: the good lies not in the sensible world, but beyond it in the metaphysical world.

The morally good, which appears as celestial bliss in various religions, is henceforth separated from what appears good only on earth, or earthly happiness. The pleasant does not coincide with the good; in fact, the morally good is usually assumed to be devoid of pleasure. The concept of value is made absolute and hypostatized in various ways. We find that the aspirations ascribed to man in the hereafter are the same as his aspirations here on earth, that his happiness there is just like the happiness he desires here.

There are further attempts to make the concept of value absolute following Plato's example. Duty and inclination, happiness and virtue are sharply divided from one another; what one would like to do is now universally distinguished from what one ought to do. The concept of value, as used in antiquity, has to be modified for this purpose. As with most philosophical problems, we see in this case *how an old word is given a new sense but that this new sense is not taken into account*. The so-called 'absolute value' is a concept constructed by philosophers. In antiquity they asked: How must I act *in order to be happy*? (They did not consider that one might want something other than to be happy.) In modern times they asked: How ought I to act? They did not add the above condition of being happy and thought that without it they were already asking something that made sense. This completes the process of making the concept of value absolute; the morally good, considered as what is valuable in itself, is completely separated from what man strives for. This is also what distinguishes modern from ancient ethics: the latter does not recognize this kind of moral conduct which would be valuable in itself. There is no justification to be found in Kant for making this concept absolute; though it was Kant who took these ideas, which made sense when they were part of the language of everyday life in antiquity, and reduced them to a philosophical formula. In its absolute form the concept is completely metaphysical, for there is no absolute value to be found in the world.

The theory of value is henceforth pursued as a special part of metaphysics beside the theoretical part. The realm of values is something metaphysical that exists beside the so-called genuine reality. The basic idea here is that there is a value independent of our wishes. However, what is to be understood by this is not considered carefully enough and is instead assumed to be known. Those who defend this idea again resort to a trick. If a philosopher asks for a more precise definition, he is blamed not only intellectually but also morally. He is told that he is 'value-blind' and immoral because he does not know what it really means to be moral. This kind of problem, which arises in modern ethics, has no parallel in antiquity. It is a product of modern civilized man.

The question whether real moral conduct is opposed to one's wants and wishes is connected with the question whether such conduct is possible. This question, i.e. 'Can man do what he ought to do?', is called the problem of free will. There is no way to formulate it clearly. Philosophers generally ask: Is our will free or is it determined? It looks as if morality necessarily presupposed that action is undetermined. The word 'freedom' has been used to characterize such a state. Man is supposed to be responsible for what he does only if he is in such a state. And yet, it is said, human character is determined by various circumstances which all 'obey' the principle of causality; so actions too obey the principle of causality and are therefore unfree. Such 'problems' are still presented today as the main problems of ethics. After critical examination, we shall find (1) that the formulation of the question is completely wrong, i.e. one cannot tell from the question what is being asked. And we shall see (2) that the question whether the principle of causality is universally valid (determinism versus indeterminism) has nothing to do with ethics.

We see that the problems of ethics point in the same direction as those of theoretical metaphysics. In fact, Kant finds that the only way to arrive at metaphysics is to start with ethics. Ethics is to him a new way to enter an otherwise impenetrable thicket. In ethics we have a confused mass of undecidable questions and answers, just as in discussions of the nature of being.

The concept of value has also been given an aesthetic form. In aesthetics we distinguish a school which tries to give a more empirical answer to the question 'What is the beautiful?' and therefore proceeds psychologically. There is another school which says that the truly beautiful is *independent* of what is felt to be beautiful. There is a similar situation in the philosophy of religion with respect to the holy, etc. For purposes of criticism, we shall first take up ethical problems. Using their clarification as an example, we shall also become clear about the problems of aesthetics.

JOSEF SCHÄCHTER

THE MEANING OF PESSIMISTIC SENTENCES

JOSEF SCHÄCHTER

THE MEANING OF PESSIMISTIC SENTENCES

'Vanity of vanities, saith the Preacher, vanity of vanities; all is vanity' (*Ecclesiastes* 1:2).

'Yea, better is he than both they (the dead who are already dead and the living who are yet alive) who hath not yet been, who hath not seen the evil work that is done under the sun' (*Ecclesiastes* 4:3).

'Never to have been born is much the best' (Sophocles, *Oedipus at Colonus* 1224).

'And there is no new thing under the sun' (*Ecclesiastes* 1:9)

These pessimistic sentences, which seem at first sight easy to understand, turn out on closer inspection to be meaningless strings of words. Our analysis of them shows that we are dealing with combinations of words which have no meaning and cannot therefore be called statements. To describe how such an investigation is carried out, we divide it into three stages.

(1) The first stage is that of naive understanding perfect in itself. At this stage such sentences seem to us very familiar – especially when we are, or at least have been, in a bad mood. We understand them and grasp their content without hesitation.

(2) The first stage is followed by the stage of logical analysis, which brings out that these pessimistic sentences are merely meaningless combinations of words.

(3) This is followed by a third stage in which we look around for a solution, troubled by the incompatibility between the familiarity and true-to-life feeling of those sentences and their logical meaninglessness. There are two solutions worth considering: (a) the symptomatic one and (b) the one based on my distinction between different 'language strata'. (The general logical aspect of this distinction will be investigated on another occasion.) In this essay I shall first deal with the second stage and using simple means try to make the reader see that these sentences are meaningless, before

going on to sketch the solutions and in particular the one involving language strata.

Concerning the first sentence, the one taken from *Ecclesiastes*, it should be noted that if a human action is to be *evaluated* from a certain point of view, it must be aimed at a purpose. When an action is described as 'good' or 'evil', 'useful' or 'harmful', 'important' or 'vain', this is done primarily with a view to the purpose which the action in question is trying to achieve. Strictly speaking, purposive and only purposive actions are measured by the yardstick of value. When I go to the library in order to read a certain essay, my taking this way or my going there can be considered valuable or condemned as vain, depending on whether reading the essay is regarded as important or unimportant. Usually such value judgments are not about isolated actions, but about many interconnected actions all aimed at a *single* goal and hence trying to achieve a *single* purpose. Thus students attend lectures every day, take notes, read, ask questions and discuss, etc., in order to go into a science or in order to further it. We have here a whole system of actions serving a single purpose, i.e. the one just mentioned. We act over a fairly long period of time in the service of a *telos* that dominates everything we do. Purposes too are often interconnected and form *systems of purposes*. This simple account is only about *conscious* actions and *conscious* purposes; the question whether there are unconscious purposes and what they are like will here be disregarded. From this point of view, a person's actions over a certain stretch of time can be judged positively or negatively only with a view to the purposes he has been pursuing. Natural events cannot be evaluated; they cannot be praised or blamed, where praise and blame are taken to indicate 'genuine' value. Natural events are evaluated and divided into positive and negative only in a metaphorical sense, according to whether we welcome them or curse them. More on this below. For now, we would stress that only conscious purposive actions can be 'good' or 'evil', 'important' or 'vain' in the proper sense. This statement corresponds undoubtedly to the rules implicit in our language. However, the concept 'all' in 'All is vain' also comprises all natural events and sequences of them: the sun that shines, the wind that blows, the rivers that run, etc. (See the preacher's subsequent sentences in *Ecclesiastes* 1:2 ff.) But natural events cannot be vain because they are not purposeful actions. Thus regarded at least from the linguistic point of view (i.e. that of the rules governing language), the preacher's first principle is not entirely unobjectionable.

Another objection seems to be more powerful: If someone tells us that the world as a whole or human life as it is at present is vain, he would have to be able – at least theoretically – to specify a state in which the world or life was *important*. Then and only then would the assertion of vanity have a meaning. If the preacher were to tell us for example that as soon as the 'eternal monotony' ceased, as soon as the sun shone brighter every day or every year, or injustice vanished from the world, or death and misfortune ceased to afflict us, he would under these circumstances give up his thesis about vanity, then this thesis would have a meaning. But since nothing is apparently further from his mind than restricting his thesis to certain empirically given situations, and since he wants it to be taken instead in the *most general* sense, he thereby robs it of all meaning.

The outward form of pessimistic sentences is deceptive. For these sentences look at first as if the despair expressed in them was about a given state; but such pessimism would not be *genuine* because it tacitly presupposes that another possible (but, alas, unreal) state would be valuable and satisfactory. This is why contempt for worldly goods is genuine only in people like the preacher (see *Ecclesiastes* 2) who have plenty of opportunity to enjoy them. For the pessimism of poor people is, strictly speaking, an affirmation of the value of earthly goods. We can even go so far as to say that *the only genuine pessimism is one which regards no imaginable state of the world as valuable.*

Anyone who speaks of the vanity of an object or activity must also be able to specify circumstances under which he would call the object or activity important and valuable. If he cannot do this, his statement is meaningless. The fatal difficulty in the vanity thesis is that it is meaningful only if one can specify a state in which everything would be valuable; but if one cites this state, the pessimistic thesis loses its real meaning and turns into a program, an idea striving to be realized and as such opposed to the vanity thesis. As soon as someone describes the valuable state, it appears desirable and important and the pessimistic attitude is overcome. I call this situation 'antinomic tension' and this observation *the most important antinomy of pessimism.*

The sentence 'It is better for one not to have been born' seems open to an even simpler objection. The meaning of the complex sign 'state a is better for x than state b' is meaningful if and only if x exists in both state a and state b. But if x cannot exist in state a – as is the case with the preacher's and Sophocles' sentences, for x's existence in state *a* has been explicitly excluded – then it is logically impossible to say that state *a* is better for *x* than state *b*. This observation constitutes a further antinomy. The pessimist's denial of any progress is also antinomic. If someone denies any human progress (as the preacher does with the sentence: 'There is no new thing under the sun'), we ask him what he understands by 'progress'. If he means technological progress, then he is advancing an historical hypothesis about the presence of technological achievements in ancient times, which is indeed logically unobjectionable, but must first be proved. In principle, one can neither affirm nor deny this conjecture, but must simply await the results of new excavations which could make the hypothesis probable or improbable. In any case, in the present state of archaeology and prehistory we have no reason to take the hypothesis seriously. On the other hand, if by 'progress' he means improvement of moral and social conditions (as is very often the case with such theses), we ask, e.g., whether abolition of slavery does not mean moral progress: if he answers this or similar questions in the affirmative, he accepts moral progress. But if he defines moral progress in such a way that by it he understands a state which has not yet occurred (such as an end to war, among other things), he must be able to describe this state; and as soon as we have such a description, his denial of progress means the conjecture that such a state will never occur. This conjecture may, of course, prove true, but it may just as well turn out false, like any other prediction about the future; and we must therefore await future states before we can decide the truth or falsity of such a conjecture. However, we can at least consider indications that are already present and tend to show that there is no progress. Such a conjecture would, of course, have to be further amplified by stating on what observations and laws it was based, whether these seemed well enough corroborated and by what, etc. It is easy to see that this conjecture does not mean the same as the thesis denying any progress; such a conjecture could not for example be expressed in absolute and universal terms. All one could say is that one can conclude from certain symptoms with a certain degree of probability that progress is not to be expected in the field of ethics (where this term remains to be defined with more precision).

A similar fate awaits Schopenhauer's well-known view that the highest degree of pleasure consists in the absence of pain. The sense of this assertion seems to us very problematic if we consider that it makes no sense to speak of pain if one cannot specify the opposite state (that of pleasure). But according to Schopenhauer, it is impossible to specify the state of pleasure, for this specification has been explicitly ruled out by definition. This shows that the sense of this sentence can also be called into question.

The arguments mentioned incline one to regard utterances of this kind only from the *symptomatic* point of view, i.e. to regard such a thesis not as a statement (like those of everyday life or science) describing some state of affairs in the world, but as an indication that the utterer must be in a certain mood. It is thought that the difficulties can be avoided in this way. For a sentence may be very interesting and revealing when viewed under its symptomatic aspect and yet fail to represent any state of affairs in reality. (See my *Prolegomena*, part I, ch. 1, on 'Symbol and Symptom: Two Aspects'.)

At first this solution may leave us only with a feeling of dissatisfaction; but the feeling is so clear and distinct that we cannot easily ignore it. How is it possible, we ask, for sentences like the ones discussed to seem so plausible, genuine, and true to life? How is it possible for such 'meaningless strings of words' to carry such conviction? It seems paradoxical to us that those combinations of words, which are really only pseudo-sentences, should play such a big part in our lives and actions. And if we look more deeply into language, this feeling seems to us justified; this is why we will now try to indicate the place that pessimistic sentences occupy in language.

It is important to note that our language consists of various 'strata' between which there are various 'correspondences'. I regard this as an important insight, which I am only adumbrating here. (I try to define the strata and the correspondences more precisely in a paper devoted to this thesis.)

We can, for example, distinguish between a language stratum which is neutral with respect to 'values' and another in which all things and processes are measured by a yardstick of value. Within the value-free (neutral) language we distinguish, e.g., a grammatical and an applied language (see my *Prolegomena*, part I, ch. 2); the same holds for the language of values, as language strata need to be differentiated according to aspect.

In the analysis of language, we come down to a stratum where all events are regarded only from the human point of view, where the sun, for example, was created for the purpose of lighting up the earth and thus in order to serve mankind well. By this I mean not only that these views were prevalent at one time, but also that they have shaped language, i.e. led to the laying-down of certain rules which still hold in our language (or more precisely, in one of its strata).

This is very important for understanding our method of investigation. We do not want to establish new hypotheses about earlier strata of human thought (e.g. about its animistic or magical stage) nor about modern man's 'unconscious'; what we want to do is uncover rules already present in our language. This is not to deny that these 'deposits' of rules in our language go back genetically to earlier stages of development and that these stages of development are also present as components in our thoughts and feelings; it only means that in our investigations we will be looking at strata exclusively under their linguistic aspect.

An individual human being appears in one of these 'strata' as having always existed and existing for ever; the individual 'I' is dominant. This view or picture of the world leads for example to the linguistic rule that the question 'Why?' can also be asked about natural events even though they are value- and purpose-free when looked at from the point of view of natural science. The fact that such a language stratum is also present in our everyday language is shown by the usual forms of expression, such as 'The weather is good', 'It was a bad winter' and the like. Anyone who takes the view that these are fixed turns of phrase or casual figures of speech should remember his own omissions and commissions and the thoughts and images that govern life and action and have not yet been subjected to intellectual control. The way we behave in company often differs from the way we behave at home, and mixing the two spheres leads to confusion. In our analogy, the thoughts and modes of expression controlled by intellect correspond to our behaviour in company, while the original uncontrolled thoughts are like our behaviour at home. We are inclined to translate everything into 'polite' 'scientific' language and would like to call any deviation from this language a metaphor. (In company we behave in such a way as to make others believe that we behave the same way at home; a superficial observer may really think that there is no difference between the two ways, and only an experienced one may be aware of the difference.)

There are turns of phrase (combinations of linguistic signs) that are forbidden in one language stratum but perfectly legitimate in another. If a language stratum is such that the grammatical category of purpose is dominant in it, so that the question 'Why?' can also be raised with respect to natural events, and things that happen without motives and decisions are regarded under the aspect of purpose, we call it a 'teleological' language stratum, and we note that – in so far as it is internally consistent – it *cannot be refuted*. For a language stratum is constituted by its own rules, and rules are neither true nor false. These rules constitute the logic peculiar to this language stratum.

If the rules agree with one another, the language stratum has, logically speaking, been correctly formed, and there is nothing more to be said about it from the point of view of logical grammar. All one can do is refute the *empirical reasons* that led to the formation of the language stratum by making use of other empirical knowledge (e.g. by showing that those items of empirical knowledge that appear to be the reasons in this context are improbable) and go on to suggest the *inappropriateness* of the language strata formed for those *'reasons'*. To elucidate this using our example: the teleological language stratum came into being by reason of certain explanations of experience (animistic, magical or others); later on, natural scientists explained those experiences in another way without reducing them to purposes and intentions. It turned out that the scientific explanations were more plausible and probable than the original explanations, so that it was possible *to overcome the reasons for forming the teleological language stratum*. A scientist who succeeds in getting us to abandon old explanations and accept new ones may sometimes also succeed in getting us to be disloyal to our old language stratum and thus help a new one to win our loyalty. (Of course, old, abandoned gods often 'avenge' themselves and regain their power over us in our 'weak moments'.) Abandoning a language stratum means transvaluating former values, and like any valuational act, it is, strictly speaking, independent of experience (i.e. of the sentences that describe it and of the truth and falsity of those sentences). Only the reasons leading to the formation of a language stratum can be overcome by experience and, in particular, by new attempts to explain experience. Thus a self-contained language stratum can be overcome only indirectly.

There is another language stratum much more difficult to characterize than the one just mentioned, a stratum in which certain concepts and grammatical categories display a curious anomaly, which makes it more difficult to describe. I should like to call this the *ideal* language stratum, and I will try to elucidate it by means of a grammatical category which assumes a different meaning in it, namely the concept of 'value'. What is regarded as valuable in this area is not *things* and *states*, but *'directions'*.

The word 'direction' is meant figuratively in this connection; for example, a stone can be thrown in two ways: (1) we see the target clearly and try to hit it; (2) we merely guess that there is a target there and throw in that *direction*, but without seeing it clearly. In the same way, our striving for 'absolute' correctness corresponds to aiming at a target which is only known to lie in a certain direction.

Let us elucidate this by an example. Someone has heard many nice things about the Alps and idealized what he has heard; not being in a position to visit any of the Alpine countries, he has made himself an even more ideal picture of the Alps. Finally, by a stroke of luck he is able to travel to the Alps, but what he sees does not live up to his ideal. And yet he is unable to describe the desired state. Are we now to call the optative sentences describing his original expectations meaningless because he cannot specify the hoped-for state? There are many sentences in our language which are largely indeterminate and yet cannot be characterized as meaningless. Man just entertains some wishes without being able to state what exactly their fulfilment looks like. In our culture, the concept of 'value' as such has also been *idealized* in this way. From the perspective of the rest of language, to idealize value terminology is, in every case, *to go beyond* the limits of language. 'Value' is no longer a term designating the sets of concepts 'good – evil', 'beautiful – ugly' etc., where it is possible to state how each element is realized (applied). The question 'What is valuable?' or 'What does a state (or action) that has value look like?' is not answered by *describing* a state of affairs, but by pointing in the *direction* of a wish.

In our mental life we sometimes apply types of sentences which resemble other familiar types of sentences but are not to be confused with them. Looked at grammatically, daydreams for example do not consist of assertoric sentences. If we ask someone about the grammatical characteristics of his ideas (and hence 'sentences') in a daydream, e.g. whether they are true or false (or better, whether he is guided in the rest of his life by the states of affairs that appear to him in his daydream), he will usually reply – if he has understood our question – that as sentences these ideas are not to be regarded under the aspect of truth and falsity. If we then ask him whether these are optative sentences, he will also give us a negative answer – once the question has been sufficiently explained to him. This obviously suggests that we are dealing with another type of sentence which is neither assertoric nor optative. But for now this should be regarded as a mere suggestion.

There are sentences in our language which are expressed inside a language stratum and are meant to reject the language stratum (in which they are expressed). Such sentences, which must, as it were, go against the rules by definition, may be characterized as *'resignation sentences'*, i.e. as *sentences in which someone who resigns from a certain language announces his resignation in that language*. Among such sentences are the preacher's vanity thesis and the equation of pleasure with reduction of pain (à la Schopenhauer). 'All is vain' then means: You have felt compelled to form the set of concepts 'important – vain'; I reject this language system! But this 'sentence' ('All is vain') can also be interpreted to mean an ideal unimaginable state, an indescribable direction of a wish, which does not imply a rejection of that set of values; it then means: everything is vain in the 'direction' of my wish. Similarly, 'All is pain' means either: I reject your language containing the pair of concepts 'pleasure – pain'; or: Compared to my unrealizable pleasure, any ordinary pleasure is really pain. The denial of progress etc. is also to be understood in one of the two senses mentioned.

If we analyse our use of the personal pronoun 'I', we find – apart from other epistemologically relevant meanings – two diametrically opposed ways of using it: (1) 'I' may mean: Mr. N.N., who was born in year a, has since then undergone certain changes and will no longer be alive after a – not precisely determinable – number of years. This is, as it were, the *neutral* (scientific) meaning not infiltrated by values and wishes. It is an 'I' defined only in terms of *assertoric sentences*, without any sentences expressing values or wishes. But the word 'I' is also used in another way: so as to include our hopes and value judgments. This 'I' is defined in the *'language stratum of immortality'*. It does not begin with year a and end with year z. In the sentence 'It is better for one not to have been born', the subject exists before his birth; we are therefore dealing with the latter meaning of 'I'. This meaning can be given in two ways: as a real description, i.e. as a soul which first exists in and for itself and only then enters a body, after which it leaves it again, but without therefore ceasing to exist; or as the direction of a wish which is not even realized in a description. Such concepts, which are given only by their direction, are very characteristic of modern thought. We have banished the mystical way of thinking from our world view (which consists of assertoric sentences), but without eliminating it from our *'language of interests'* (which consists of sentences expressing wishes and values).

Nothing is more characteristic of our linguistic attitude than the fact just stated.

In such strata, which have only been adumbrated here, the preacher's sentences, which were cited as sentences characteristic of pessimism, seem to be perfectly meaningful, and since these strata are part of our language and play an important part in it, it is understandable that they should stubbornly resist the label 'meaningless'. In these strata it seems perfectly legitimate to raise such questions as: 'Why does man breathe?', 'What does he live for?', 'Why does the wind blow, the sun shine, the river run, ...?' In these strata we can formulate such sentences as 'All is vain', 'It would be better for me if I had not been brought into this world', and the like. These sentences and questions are understood, for to 'understand' a sentence is to fit it into an established language stratum.

JOSEF SCHÄCHTER

NOTES ON PROBLEMS OF ETHICS AND THE PHILOSOPHY OF CULTURE

JOSEF SCHÄCHTER

NOTES ON PROBLEMS OF ETHICS AND THE PHILOSOPHY OF CULTURE
Vienna 1937

In the following sections I should like to draw attention to some fundamental points in ethics and the philosophy of culture.[1]

PSYCHOLOGISM IN ETHICS

In so far as thinkers of the Vienna Circle gave their view on ethics, it was a psychologistic one. Ethics was regarded as part of psychology. This is also the attitude underlying Moritz Schlick's *Problems of Ethics* (first published as *Fragen der Ethik*, Vienna: Springer Verlag, 1930). The law of motivation plays more or less the same role in his ethics as does the law of association in traditional psychology.

Now that psychologism is as good as overcome in logic, it is time for it to be overcome in ethics as well. From a logical point of view, it is misleading to reduce the laws of moral conduct to psychological and sociological and hence natural laws (where 'nature' is understood in the widest sense, so as to include also human behaviour and the nature of human society). The laws governing human conduct have their origin in human nature. But even if we succeeded in understanding the laws of physiology, psychology and sociology as well as we understand the laws of physics for example, this would not *justify* their *validity*. They would not be *well-founded* for that reason. A law of nature describes, explains, but does not oblige. A law of conduct does not describe, explain, but obliges. The word 'law' means something different in ethics from what it means in the sciences. A law of psychology or sociology is valid, i.e. is regarded as true, as long as we have not convinced ourselves that it is false, whereas a law of ethics is valid, i.e. obliges, until it is abolished. A law of psychology or sociology depends on observation, i.e. on whether it interprets observed data well. Whereas a moral law depends on its acceptance by the group of people to whom it applies, though in many cases a moral law is *confirmed* by the conduct of those who *transgress* it.

It is well known that inconsistent reactions and a constant need to justify one's actions are the best witnesses to the existence of a moral law in one's culture. We are faced with a paradoxical situation: *falsification contributes to verification.*

JUSTIFICATION

Since ethics consists of norms, and since norms have the characteristics of imperative sentences and are often justified, we shall have to say something about this kind of justification. Assertoric and imperative sentences are two different types of sentences. The difference is much greater than is commonly supposed. It comes out also in the ways they are justified. Take e.g. 'The slaves have brought the stones, *for* their master ordered them to bring them' and 'Bring the stones, for I, your master, order you to bring them'. The two sentences resemble one another in a sense, but differ otherwise from one another. The first describes the causal connection between the master's order and the slaves' action. The second serves to strengthen one's influence, to oblige others, to *induce* action, and not to *explain* facts by means of other facts. An imperative and an assertoric sentence are thus justified in different ways, and this difference is what matters. Ethics cannot be part of psychology any more than logic can, because *ethics like logic deals with norms.*

Any ideology is essentially normative. This is why no ideology can be justified scientifically. There is no scientific justification for rejecting such practices as cannibalism. The sphere of demands and commands differs from that of description and causal explanation. The points of contact between the two spheres are given along with the structure of human existence. These points of contact are represented linguistically as correspondences and similarities, and this is what leads us to equate the two spheres.

THE LAW OF MOTIVATION

What makes man act? Why does he act? Schlick answers this question by saying that any idea, any content of consciousness has a certain property which causes us not to be indifferent to it. This is the property of being pleasant or unpleasant, of giving pleasure or pain. Among the ideas, feelings or other mental acts that act as motives, those that prevail are the ones that promise more pleasure or less pain. They are the ones that make us act. Their pleasant or unpleasant character is called their pleasure value, and this is the motive force behind our actions. The law of motivation therefore states: If a person has to decide whether or not to perform a certain action, he always decides on whatever course promises more pleasure or less pain. If a child is presented with two pieces of cake, one of which is bigger than the other, and if he would like to have the bigger one, but 'forces himself' to take the smaller one so as to leave the bigger one to someone else, then he is acting morally. But even this action proceeds according to the law of motivation, for owing to the child's character and education, the process of motivation in him differs from the corresponding process in another child who chooses the bigger piece. The former also chooses according to what gives more pleasure, for all voluntary acts proceed according to the law of motivation. Even a martyr and a hero act according to this law. The ideas of the goals they set themselves have a pleasure value for them.

This calls for the following remarks. While this law looks like a true law, regarded critically it is either tautological or vacuous or false. A law of nature that represents real processes must be verifiable or falsifiable. A law that cannot be falsified by experience is not a law. The law of motivation is not falsifiable. For in any case in which someone decides to perform an action, the action is said to promise pleasure to him. If someone runs away from a fire, he does it in order not to burn himself and hence to avoid the reduction of pleasure due to the pain. If an ascetic deliberately sticks his hand into the fire, he does it in order to gain the pleasure value implicit in asceticism. If a child chooses the bigger piece of cake, it is because it promises more pleasure to him in the eating of it, and if he chooses the smaller piece, it is because the idea of being a good boy promises more pleasure to him. The law is right in any case, and this is no recommendation for a law of nature.

When we deliberate before reaching a decision, various motives are at work: e.g. a craving for a momentary pleasure, a striving for satisfaction or for giving meaning to one's life, etc. The decision is the result of a struggle between such motives. We act with a view to what promises pleasure, but we also act with a view to the meaning of life. And we also act with a view to doing our duty. But if what promises pleasure does not give meaning to our lives or goes against doing our duty, we decide in favour of one thing rather than the other, and this decision proceeds according to *several laws of motivation*. In order to formulate these laws, we must separate the motives from one another and then ascertain on the basis of experience which motives prevail under given circumstances. Such laws of motivation, if established by scientists (psychologists), will be verifiable or falsifiable. The mistake in the law of motivation cited above is that the word 'pleasure' (or 'pain') embraces all possible motives of our actions, with the result that we overlook the important differences between them and are no longer able to establish any real laws.

THE COMMON ELEMENT

Ethical views and norms originate in elementary needs common to all human beings. Some of the needs at the origin of ethics are also present in other living beings. As the presence of memories, wishes and goals in human consciousness has given rise in language to the categories of the past and the future, so the different positions of influential and uninfluential people, males and females, etc. in human society have led to moral categories. The moral meaning of an action is a kind of average value of the reactions of people of a certain milieu to that action, and it is thus like the meaning of a sentence which, when regarded under this social aspect, is the average value of the ways it is used; in logic, as in ethics, we are dealing with the average value of the uses and actions of certain individuals of a certain group. We are looking for the common element in the moral judgments of different cultures and periods. At first we come upon big differences: polygamy is prohibited in one place and permitted in another, etc. And yet in the final analysis we are dealing with the same items. We are dealing with higher things, with the value of human life, with loyalty, friendship and love. The forms differ, sometimes to the point of incompatibility, but the common element is clearly recognizable and can be established empirically. This kind of research will be called 'empirical ethics'.

The empirical side of ethics thus consists in finding the common element in different cultures and periods and bringing out those basic features that experience shows to be essential to any human culture. In doing this kind of empirical work, we must also take into consideration the moral conduct of living things other than man.

LOGIC AND ETHICS

If we want to understand ethics, we will do well to compare it to logic. Man is said to be a thinking being. Man could also be defined as a valuing being. Why has the former definition been preferred? Is it because animals also value things in a way?

But in a way animals also think. Is it perhaps because there is such a big difference between human and non-human thinking? But is the difference in valuation smaller?

The law of contradiction plays a big part in logic. But the same is true of ethics. A person who wants to introduce order into his thinking must not affirm and deny the same thing at the same time. And if he wants to conduct himself in an orderly fashion, he must make sure that his conduct is free from contradiction. Being true to oneself is an ethical as well as a logical criterion.

Not every use of signs is accepted into the logic of language. In popular speech double negation is used to strengthen negation. But the rule accepted into logic is that double negation means affirmation. This rule is also read off from usage. What is at work here is a kind of automatic selection. In many cases the average value of certain actions by certain people becomes law. The process of selection is complicated. Only the results of the interplay of forces become visible. Anyone who wants to understand the rules of the logic of language or of human conduct must read the book of life. This reading requires wisdom; mere routine does not suffice. For it requires the ability to distinguish what is more important from what is less so in that book.

The two fields, ethics and logic, cast light on one another. Just as the supreme laws of logic are not relative, in the sense that it is absolutely impossible for man to change them, the laws of ethics are essentially unchangeable and only change their form. If someone denies the supreme rule of logical notation, namely the law of identity (which states: If someone uses a sign, e.g. *a*, and later goes on to use this sign without changing its definition, this sign continues to have its earlier meaning),[2] then he cannot talk with others or even think by himself, and so it is if someone denies the supreme principle of ethics, namely the principle of valuation (which states: There are things to be valued, and as long as I regard them as valuable and do not change my mind on this point, they retain their value).

If one did not value anything, one could not act in any way, not even by oneself. In life as in literature people often contravene the law of contradiction. They say or think something and its opposite. But this contravention does not abolish the law of contradiction. People also often offend against values, against higher things, against the value of human life, against the value of love. But such a contravention does not abolish the supreme principles of ethics either.

There are logicians who make ideal, maximum demands on logic. They require that every word be precisely defined and that every sentence be correctly constructed. Even though these goals cannot be attained, they are nonetheless important because in striving towards them people achieve things they would not achieve otherwise. There are also maximum demands in ethics. Some people are not satisfied with social ethics. Their ideal of the right way to live would not be to lead an orderly social life. To them, society is not the highest thing there is. The value of such demands for others lies in pointing out to them that there are higher goals above the ones that appear to them to be the highest. It is a fact borne out by experience that there are people for whom turning to the highest changes their entire lives. People around them feel troubled by their presence in their midst. They display some characteristic attitudes towards them, but in no case do they remain indifferent: they either glorify them or fight and destroy them. The violence of their reactions constitutes an immediate unintentional admission on their part that these values are also present in them and important to them even if they themselves deny them. This is an important empirical fact. For indifference in such cases would testify to their lack of values. Logicians also cause offence, though less so, and for understandable reasons.

The two fields are also similar with respect to what we do not know. Just as we do not know how and why the supreme laws of logic have arisen and why just these laws have been accepted, we do not know how and why the supreme principles of ethics have arisen and been accepted. To express this in religious terms, one could say that such laws are the result of man's contact with God. To avoid such a form of expression, one could say that they have their origin in elementary living conditions common to all men. This is a terminological question (though one terminology may have a stronger influence on people than the other).

Philosophical clarification in the field of ethics is analogous to philosophical clarification in the field of logic. The classical example of it is

Socrates' conversations in Plato's dialogues. The relationship of ethics to jurisprudence, sociology, education and politics is like the relationship of logic and mathematics to physics, or like that of axiomatic to physical geometry. Here too, there is what is known in the terminology of the philosophy of nature as the 'problem of application'. More about this elsewhere.

The part of ethics comparable to pure geometry will be called 'pure ethics', while the part of ethics comparable to physical geometry will be called 'empirical ethics'.

CULTURE WITHOUT MYTH

It is essential for a culture that its members have a common conception of life and find a common meaning in life. The conception of life must convincingly explain the *important* functions and at the same time give a meaning to life. Human beings did not create themselves. They therefore want an explanation of who created them. There is something higher and more powerful than they, and they want to give expression to this higher and more powerful being. Human beings die, and they want to know something about death and what comes after death. They have relationships with their mother, children, members of the opposite sex, animals, etc. These relationships have the common characteristic of being *important*. A conception of life is correct if it explains these relationships in a satisfactory manner and at the same time gives a meaning (goal, content) to the lives of members of the culture, a meaning that transcends their lives. The test of this transcendence is that members of the group are ready to sacrifice their lives for this meaning (goal, content) and actually do so. The meaning of life is higher than life and independent of it. And this is what is meant by 'culture'.

The cultures underlying today's European and American culture were either mythical or quasi-rational religious ones.[3] (Greek philosophy was an exception; it was confined to certain circles within a culture.) Mythical cultures fulfilled the above-mentioned functions. They were characterized by great intensity, potency and vitality. But even the quasi-rational religions that succeeded the mythical cultures (the classical examples of which are the Mosaic, Christian, Muslim and Zoroastrian religions) did not lose that potency. The question arises whether that potency does not derive from the vestiges of myths within them, which would lead one to conclude that the religious cultures (and hence culture in general) will be exhausted as soon as the last vestiges of myth within them are used up. We are here taking the optimistic view that *culture does not lose its potency with the disappearance of myth.*

After the quasi-rational religions of our cultural tradition were shaken by science and technology, our culture began to crumble. Both the conception and the content of life were badly damaged. It should be noted that the shocks delivered by science and technology did *not* begin with science and technology. The first shocks were delivered by the problem of reward and punishment, which plays such a big part in the Old Testament and which is what the Book of Job is all about. The shocks merely increased greatly in severity in the most recent period.

It should be pointed out that it is impossible to renew myths, i.e. to return to the state prior to the quasi-rational religions. Nietzsche wanted to revive Greek or Germanic myths, and many thinkers of the last few generations wanted to regain the lost paradise of myths; but we must not yield to such illusions. Anyone who wants to resuscitate myths will not create a true theory. The main problem now is whether it is possible to create an atmosphere in which a strong culture is able to flourish even though the residue of myth has greatly dwindled in recent times as the religious cultures have grown weaker, and whether values will be strong enough in a world where important life functions have been neutralized, i.e. consciously regarded mainly from the scientific point of view and hence separated from important life centres.

Our considerations are based on the belief that man has innate biological regenerative powers, which are at work repairing the damage due to such disturbances as those caused by science and modern life, and which will make it possible for values to become again a powerful force in man.

OPTIMISTS AND PESSIMISTS

The optimists say: Modern man has a big stake in values. He is known to be very interested in a good social order, and in so far as he is interested in it not just for his *own* well-being but for the sake of a better world after he himself has gone, we are dealing with an expression of the will to realize the good. If we listen to young people talk, we can gather from what they say that values still occupy a relatively large place in their lives. If someone were to object that these young people reduce their values to material things, anyone trained in logic could easily convince himself that such reductions do not stand up to criticism. Their striving for liberty, equality and the like cannot be easily explained in the ways they themselves indicate (in terms of historical materialism, psychoanalysis, etc.). If they strive for pleasure, happiness, satisfaction, it is not for pleasure *as such* or happiness *as such*, but for a *certain* pleasure. They do not consider everything pleasant to be good. They therefore recognize a good that is independent of the pleasant. This 'independence' or non-identity of the pleasant and the good, or of the good and the useful, *generates culture* and is therefore of the greatest importance. Many people may well be indifferent to values. In particular, people of more mature years are often found to be indifferent to them. Many of them have given up their values and narrowed down their world to their daily lives. But anyone who looks more closely at them can often discover under the smouldering ashes a spark of what once was or must have been in them. However, we are not here talking of those people, but only of those who still lead real lives.

So much for the optimists. The pessimists on the other hand say: Even the world of the best among the moderns shows a lack characteristic of our time, namely pointlessness. Their world consists of separate spheres, the spheres of nature, morals, etc., and this is an impediment to cultural development. Unfortunately, the only thing these separate spheres have in common is their pointlessness, i.e. they are no longer vitally important. The result of the intellectual shake-up of the modern age is a pointless, senseless world. There is chaos everywhere. For what is essential to modern agnosticism is not just the attitude expressed by 'I don't know', but also that expressed by 'It's not worth knowing'. In Socrates' times people already asked: What are values based on? Why should we strive to be good? What for? They answered that the good is useful; but there is proof that evil can be even more useful. And in any case, what is the good? For what seems good to one individual or society may be evil in the eyes of another. Today too, young people raise such questions, and so do older people whose minds have not been dulled and made indifferent by the years. The defenders of values point to such important facts as that there can be no human thought or action which is not based on valuing things and does not presuppose that there are values. Values are like natural numbers or colours in that we cannot prove their existence, and if an opponent stubbornly refuses to acknowledge them, we cannot prove the contrary to him. But if you let him speak and act and you then examine his speech and action, you will find that he too makes use of numbers, colours and values. For without them we cannot think or speak or act.

But people are not satisfied with such knowledge. Values occupy more space in human life than do truth and knowledge of reality. And even if some people are able to content themselves with something like the Kantian answer concerning knowledge of reality – that is, that the world in itself, its nature and its structure remain inaccessible to us, but that we know without a doubt that there can be no human understanding without time and space and the categories – they are not satisfied with an analogous answer concerning values. People require values to be not only universal and necessary, but also absolutely evident. And they find it hard to accept that values have an even lower degree of evidence than knowledge.

The feeling that values are unfounded operates as an important factor in our lives. Many people accept tyrannies, sometimes even with satisfaction, and feel justified to themselves because they know that there are no values anyway. Why should one system that still has some moral scruples left be preferable to one that no longer has any? Anyone who seriously raises such questions is not in the situation of someone looking for theoretical clarification, but in that of a patient anxious to find a remedy. We all know that by merely declaring a disease to be psychogenic a doctor does not yet effect a cure. And the situation is such that the sick expect an effective remedy from medical science, but it cannot meet that expectation. Many doctors take refuge in expressions that profane higher things. Some want to avoid such profanation and therefore remain silent. And finally some say to themselves: Where there is no real doctor, you have to make do with home remedies.

Here is a criterion, to be used in the future, by which one can tell whether a state of culture has again been reached: As soon as the culture-generating factors have created a climate in which life has an obvious meaning and values have not only a theoretical but also a vital basis which generates and sustains collective action, as soon as they have created a climate in which values are a powerful force and life has a transcendent meaning, we shall know that we have once again a culture.

Against the objection that it is utopian to regard this as possible, it may be pointed out that in any group engaged in intense collective activity questions about the meaning of life and the foundations of values vanish by themselves. They look like obvious questions to such people. While great intensity may well be a sign of culture, it is not yet a sign of a good culture. For this, the values of the culture will also have to coincide with the higher values of all cultures, transcending any differences in form. In particular, the values of a future culture will coincide with those of the quasi-rational religious culture, as it too will set a high value on higher things, human life, love and friendship, and these values must not be reduced to pleasure or utility.

THE 'ESSENTIAL' MAN

The type of man who will generate culture in the future will not be a crowd-follower or party-joiner or the like, but the type of man who could be called 'essential' or 'Socratic'. The religious type of man was characterized by his personal relationship to God. The essential, Socratic man could become the heir to the religious type, no matter how much he differs from him.

Here in brief are a few essential features of this type. He is not to be identified with the moral man. Although morality is especially important to him, it does not exhaustively characterize him. Besides, it should be remembered that theories that reduce morality to utility, the social contract or the like – and thus try to derive morality from what is inessential – are now widely accepted.

These theories are sometimes well-intentioned in that they try to defend values. They are mostly harmless because of their lack of seriousness. The difference between the moral and the essential man can be characterized by saying that the essential man must aim at what is transcendental, which again means that values must not be reduced to what is merely useful or the like. The essential man strives to give his life a higher meaning. The concept of 'usefulness' means something different for him from what it means for other people. Something is useful for him if it enhances the *value* of life. Very often this enhancement is utopian. It is characteristic of him that there are only two alternatives for him: either his values are obvious to him, or if he doubts them, his very being is *shaken* by that doubt; he is incapable of methodological, academic doubt.

In a time of tyranny, the essential man will not under any circumstances inwardly accept tyranny, and this is what distinguishes him from the masses, as well as from the intellectuals who are impressed by the ability of such a system to control reality, which they themselves are unable to do despite all their efforts because they never get around to carrying out their resolutions. Even when presented with a proof that population growth and rising demand rule out all solutions except a forcible one, the essential man will continue to reject them absolutely. For to him the economic world order is not an ultimate goal for which he would be willing to sacrifice his values. This argument will convince primarily those who feel that the foundations of their world have been shaken.

The essential man rejects careerism, intellectual poses and unnatural attitudes, because such attitudes do not give meaning to life. And if he nevertheless succumbs to them, he feels he has sunk inwardly, for he feels that his life has now less meaning. But if he frees himself again from them, he undergoes an inner renewal.

As mentioned before, being essential and being strong are not opposites. An essential man shows his strength by placing the meaning of life above life itself.

The classical example is Socrates' attitude to the thirty tyrants when they wanted to send him to Salamis to bring back a man who was to be put to death, so that Socrates would share the responsibility, and he refused, scornful of the danger to his life. (See Plato's *Apology*.)

Some people are able to express the idea of an essential man in literature, while others embody the idea in their lives. Only the latter are essential men; the former participate as it were in the idea of the essential, but only as long as they remain conscious of their subsidiary function; and they remain useful only as long as they do not think more highly of their logical, rhetorical or literary abilities than these are worth. But the man we are seeking is the one who lives by that idea.

We have called the essential man 'Socratic'. This is because unlike the best intellectuals (logicians, epistemologists, scientists) of the last few decades he does not confine himself to examining the concepts we use, but also examines his *actions*. The criterion of verification consists in *putting one's ideas into practice*. Only a philosophical life gives meaning to philosophy.

NOTES

[1] This essay has not been previously published.
[2] This is the presupposition of the use of any signs at all.
[3] They were quasi-rational because their view of the world contained a minimum of myth and their concept of God tended towards abstraction.

FRIEDRICH WAISMANN

ETHICS AND SCIENCE

FRIEDRICH WAISMANN

ETHICS AND SCIENCE

PRELIMINARY REMARK

The present lecture appears to date from the year 1938 or 1939, since it is headed (Ethik und Wissenschaft. Von Friedrich Waismann, Cambridge). Our translation follows the text of the German edition which was based as closely as possible on the typescript (J.21 of the Waismann Papers in the Bodleian Library, Oxford) which seems to cover two-thirds of the intended work. The rest was reconstructed from a shorthand manuscript (J.20 in the Bodleian). The difficult task of deciphering this was undertaken by Dr. Wolfgang Grassl, who also provided references for some of the quotations. We thank him for these labours, and also our translator for searching out the standard English translations for these (when they exist).

Material derived from the manuscript is enclosed in double square brackets: [[ ]]. Conjectures, uncertain readings, and completions in single square brackets: [[ ]]. '[...]' means that an expression could not be deciphered. 'EDD.' and 'TR.' have been added to the editors' and translator's notes to distinguish them from Waismann's own. – *The Editors*

What is the relation between science and ethics? Does science break up the foundations of religion and lead to or encourage immoral and unscrupulous behaviour? Is the profound crisis of today's European ethics to be laid to the charge of the scientific discoveries of the nineteenth century? It can hardly be denied that certain ideologies, like the survival of the fittest, the glorification of the blond beast and Nietzsche's idea of breeding supermen, can be traced back to Darwin. Strindberg spoke in this sense of the 'zoological world view' of our time. In a similar vein, Tolstoy condemned the whole of European culture. Must we conclude that the scientific spirit destroys ethics?

At this time when so much is uncertain, when different ethical ideals fight for supremacy in mind's minds, while the scientific spirit moves on seemingly unconcerned like a silent star, it is more than ever necessary to give an account of the mutual relations of these two powers.

Originally, each ethical system was probably connected with the ideas of a certain religion, and its precepts were ascribed to a divine origin.

For whatever reason, the religious world gradually broke up, and this is when the problem of ethics first emerged. As long as men believed in the naive sense, they saw no problem in ethics. Socrates, the first great moral philosopher of world history, was a contemporary of the sophists. Between him and the great philosophers of the seventh and sixth centuries, an important event had occurred in Greek intellectual history: the death of traditional religious belief, at least among the educated. Only then was ethics *felt* to be problematic. Pascal and Bayle, both of them deep thinkers, owed their depth precisely to the collision between belief and unbelief; just as Dostoyevsky's spirituality was rooted in this conflict. Kierkegaard's case is also very interesting in this respect. It thus seems that seeing the existence of ethics as a problem requires a complicated mental set, a certain historical position between two ages.

Thus the loosening of the theological substratum made ethics *questionable* and raised it to the level of a *problem* to be tackled by the thinking mind.

Thinkers then tried to ground ethics somehow, to give it some support. And now the great significance attached to ethics and the awe that had always been felt before it found expression in attempts to give ethics a metaphysical foundation: it ceased to be a revelation out of the mouth of a divinity and became an expression of the real meaning of existence. We thus see a series of great philosophers attempting to link up ethics with metaphysics. One example is Kant, for whom ethics was the field in which man rises from the sphere of the empirical world subject to natural causes into an intelligible world of free spirits. It is well known how these ways of thinking show up in Spinoza and determine the whole outline of his metaphysics. Schopenhauer says very clearly:

For just as every religion on earth by prescribing morality does not leave it at that, but gives it support in a body of dogmas whose chief purpose is, precisely, to support, so in philosophy the ethical foundation, whatever it be, must itself have its point of support and prop in some system of metaphysics, that is, in the given explanation of the world and existence generally. For the ultimate and true explanation of the inner nature of the totality of things must necessarily be closely connected with that concerning the ethical significance of human conduct.[1]

I should like to say that history is now repeating itself at a higher level: after breaking up that solid, sound religiousness, the scientific spirit is undermining and gradually breaking up this metaphysical pseudo-world.[2] It begins by trying to test the arguments of those theories for their soundness and ends by discovering illusions arising from a careless use of language. We are now living in this period, and if I am not mistaken, this is the basis of the extraordinary interest in ethics at the present time.

After talking about the religious and metaphysical phases of the justification of ethics, we come to the question: Can ethics be justified scientifically? Let us first come to an understanding about the question. What does it mean to justify ethics scientifically? When would we say that we had justified it? The answer seems to be: when we can demonstrate the legitimacy of an ethic in an objective manner, i.e. in such a way that any human being must recognize this ethic as binding. Can this problem be solved?

Doubts will soon arise within us. These doubts feed on the fact that in today's world there are quite evidently different political, legal and ethical ideals – it seems to me that these groups can be separated only very artificially – and so far no one has succeeded in eliminating the differences between them scientifically. Moral philosophers do often speak of *the* good and have a lot to say about the relationship between values, but they part company as soon as they have to decide in a concrete case what conduct, what attitude, or what type of man ethics demands.

But we must put this question more in terms of basic principles. In the writings of philosophers who have dealt with ethical questions we find three kinds of sentences:

(1) definitions,
(2) empirical statements,
(3) norms and value judgments.

The first step towards clarification is to notice that sentences of type (3) can never be derived from sentences of types (1) and (2). You can, e.g., give a scientific description of how men in fact evaluate things and – perhaps – find out the conditions under which certain evaluations arise or change – in this way you will get history or sociology, but you will never find out what you should do in a particular case, or what is really good.

We do sometimes find arguments which make it look as if a value judgment could be justified by an appeal to facts. We say for example: killing is wrong; *for* killing would threaten the continued existence of the human race. Here we see at once that there is a gap in the argument; it presupposes that the continued existence of mankind is good, desirable, valuable. Fully articulated, the inference therefore reads: killing threatens the continued existence of mankind; mankind's continued existence is desirable; therefore killing is reprehensible. The value judgment to be derived thus presupposes another value judgment. And this is so in every case where a value judgment is allegedly derived from facts. When Wundt, e.g., says that what is valuable is what serves progress or contributes to the creation of the spiritual goods of art and science, the circle is obvious; for what is to count as progress, as a positive development as opposed to a negative one, or what is a spiritual good, is obviously determined by a yardstick of value and cannot therefore be used to determine one. We see that a statement of value, e.g. 'That is good' or 'That is meritorious', can never be derived from a *fact*. Rather, any such derivation makes use overtly or tacitly of a further premise saying that something has value, so that we go fruitlessly in circles looking for a justification of a value judgment.

As soon as we see that value judgments can be reduced only to value judgments, never to facts, our next thought is: to ground existing value judgments on the smallest possible number of value judgments, which will form, as it were, the deductive basis of the whole system. Once the truth of the fundamental value judgments, the axioms of the science of values, is admitted, they will yield all the rest with logical necessity.

The foundations of ethics thus seem to acquire the same characteristics as the foundations of mathematics, where the process of grounding also seems to involve two steps: first, reducing mathematical propositions to the smallest possible number of axioms from which they can be derived deductively; and secondly, demonstrating the truth of the axioms. And we find that a numerous and respected school of philosophers has indeed taken this line. On their view, the problem can be solved only by showing that certain norms, the basic norms, are justified, true and adequate. And all other value judgments would be justified only in so far as they could be reduced to those initial value statements.

The whole problem would accordingly reduce to the question: Is there some means of recognizing value ascriptions, precepts or norms as right? In other words, is there knowledge of what is morally good? The answer seems to be very easy: man acts and makes decisions that are evaluated morally. When he acts, he must surely know what is right and wrong. Thus it seems that all we need to do is to analyse man's consciousness to discover in it the source of all moral judgments.

And this is in fact the opinion of a school of philosophers who may be called intuitionists. According to them, the human mind has the capacity to ascertain the presence of a value, to catch sight of the value or to get hold of it, or whatever the figure of speech may be. Some believe that values form an independently existing realm like Plato's world of ideas and can be seen with the mind's eye; others reject this hypostatization, but maintain that there is a knowledge of values that is perfectly objective and far from arbitrary. All of them agree that there is a peculiar kind of evidence indicative of genuine values. This evidence is compared and equated, sometimes with mathematical self-evidence, sometimes with the evidence of the senses.

Max Scheler for example writes that good and evil are 'clearly sensible material values *sui generis*. Of course, all this is indefinable, as are all ultimate value phenomena. We can only ask the reader to look carefully at what he experiences immediately when he feels something good or evil.'[3] Scheler continues:

A mental act of feeling, preferring, loving or hating has its own *a priori* content, which is as independent of inductive experience as are the pure laws of thought. And in either case, there is contemplation of the essence of acts and their matter, foundation and connections. And in either case there is "evidence" and absolute precision of phenomenological cognition.[4]

Similar remarks are to be found in Otto Kraus:

Sentences that do not tell us what there is in the world need not therefore "say nothing"; they already say a great deal when they tell us what there cannot be in the world. And this is what all sentences of geometry and arithmetic teach us, namely, e.g., that there cannot be a thing in the world which, when added to a thing different from it, does not yield a set of two things, or that a triangle the sum of whose angles does not equal two right angles is an impossibility. But similarly, some value axioms tell us, e.g., that an ascription of value to pain cannot be correct unless it implies rejection (or hatred).[5]

Dietrich von Hildebrand writes:

Impartial investigation shows that there is value cognition peculiar to the realm of values in which objective values founded in things and persons are given to us in a way analogous to the way colours are given to us in seeing, sounds in hearing, and things in outer perception. That we are here dealing with an intuitive grasp is beyond question.

What are we to make of such doctrines? Is there really a kind of evidence telling us with absolute certainty where there is a value? The appeal to self-evidence as the bedrock of knowledge must today arouse some distrust in any honest thinker. There is no field in which more significance has been attached to self-evidence than logic and mathematics. But it is precisely in this field that what were supposed to be absolutely evident truths were shown by more acute [[logical]] criticism to be really errors. Here are a few examples:

It seems evident (1) that the whole is greater than the part. And it seems evident (2) that two sets have the same extension if it is possible to map their elements one-one, i.e. in such a way that to each element of the first set there corresponds an element of the second set, and to each element of the second an element of the first. And yet these two propositions, which seem so evident, so obvious to us, lead immediately to a logical contraction. If we map, e.g., two line segments of unequal length onto one another, then to each point on the one line segment there corresponds exactly one point on the other line segment, and conversely; no point comes away empty-handed, as it were. We should therefore have to say that the two line segments contained an equal number of points – and yet the one line segment is obviously a genuine part of the other.

It seems to be evident that a continuous curve progresses in a definite direction at any point, and yet Weierstrass surprised the mathematical world with the discovery of a curve that is continuous at every point and yet does not have a definite direction at any point.

It seems to be evident that each predicate has a corresponding extension: the class of all things that fall under the predicate. And yet this seemingly evident principle leads to antinomies, as Russell has shown in a particular instance.

What do you make of the following proposition: It is possible to cover the entire infinite plane with squares of decreasing size in such a way (1) that no area of the plane, however small, remains uncovered by these squares and (2) that the sum of the surfaces of all the squares used to cover the plane is as small as you like, e.g. smaller than 1 mm^2? You will say: Nonsense, that is impossible! And yet this is a fact familiar to all mathematicians, which was first pointed out by Emile Borel.

In view of such experiences it is easy to understand why anyone familiar with the exact sciences is today extremely cautious if he hears someone defend an assertion by referring to self-evidence. If illusions of evidence occur even in geometry and logic, where the force of evidence seems to reach its highest degree, what are we to make of evidence in the field of value ascription, where people's convictions differ so notoriously?

However, let us not judge too hastily. Perhaps there is an intuition, an inner contemplation, which makes the difference between value and disvalue obvious. Perhaps those philosophers who appeal to such an intuitive source of morality are right after all.

Those philosophers assure us that human beings agree in their judgments of value and disvalue, even though they sometimes make mistakes and some individual human beings may even be value-blind. Is this a correct description of the situation in the field of moral value ascription? If an observer from another planet came down to earth and carefully observed human beings – I mean not only their words, but also their actions and feelings – would he agree with this report? Would he discover even *one* field of value ascription where all men were in agreement? Take a precept like 'Thou shalt not kill'. Is this generally accepted? No, for killing in war is permitted, after all. But wait, is this not just paying lip service, deferring to the power of the state, rather than expressing an inner conviction? Take another case. If there were a human monster, an extreme danger to the public, a butcher of men in the style of Genghis Khan, possessed at the same time of a formidable intelligence that made him superior to his contemporaries – would not many people declare in all sincerity that eliminating such a monster would only be rendering a service? And would not other people regard killing even *this* man as a sin? Think of the interminable discussions about whether the state has the right to exact the death penalty? What about the question whether a person suffering from an incurable disease may be killed at his own request? Is suicide immoral? Think of Weininger who committed suicide because he felt that the criminal elements in his nature were gaining the upper hand – who thus committed suicide *because* he loved what was ethical – and now ask whether it was unethical?

But, you will say, these are only exceptions, and they are exceptions only because the motive in these cases is to reduce suffering. Reducing suffering is the true ethical law we all accept. But the first thing to be said is that many people think that suffering brings depth and is therefore valuable, and secondly, it may be asked which is to be regarded as more valuable: an even-tempered life, with minor ups and downs, or a deeply moved life, up one minute and down the next.

But is not increasing happiness universally regarded as good? Stop! The concept of happiness is itself dependent on moral imperatives and coloured by an ethical ideal. For happiness is something more complicated than the mere feeling of well-being or present enjoyment. Think of how different 'happiness' looks, depending on whether the prevailing ideal is hedonistic, heroic or ascetic. The formula 'Good is what increases happiness' says nothing, as long as there is no detailed explanation of what happiness and increasing happiness is supposed to be; and this is defined differently by adherents of different ethical systems. I now ask again: is there a scientific way of deciding which happiness is true, valuable happiness? Is happiness the happiness of a warrior, of combat and victory? Is happiness mortification of the flesh, examination of conscience, and the search for God? Is happiness a clear conscience, the feeling of duty fulfilled? Is happiness *ataraxia*, a great inner calm, as the Stoics thought it was? Is happiness love? Is happiness the cheerful contentedness of innocence, idyllic contentment? Is happiness a Dionysian frenzy in which the world is filled with a divine light and we feel at one with the universe and hear the music of the spheres? Is happiness one of those rare quarters of an hour in which we are filled with a light, weightless cheerfulness, a happiness that comes on tiptoe? Or is happiness the happiness of creation? If God were to say to the soul: 'Choose the kind of life and the form of happiness you want!', how would the soul know which happiness was best?

Now suppose we say: 'Good is what leads to increased happiness in one of these senses' and think of happiness as defined in terms of a disjunction of these descriptions. But this would again be of no help; for the ideals fight one another, and anyone who sees happiness in contentment and in an idyllic life cannot preach the happiness of combat; what is good in *his* sense is evil in the other's sense.

'Good is what increases happiness.' Yes, all men seem to agree on this, but only on the wording, not the meaning, and if we go back to what they mean by the words, the seeming unanimity vanishes.

Let us consider another example.[6] If some people will not let the community benefit from their special abilities unless they receive special remuneration from it, and if they wish to use part of that remuneration to secure special advantages for their own children, then this desire is perfectly clear and understandable. If other people want all children to have the same rights, and if they wish to make sure that no one is born with special advantages or disadvantages, then this desire is also perfectly clear and understandable, even though it is opposed to the first. But what would these desires gain by each group claiming that its desire corresponds to *justice*? Is inheriting private property moral or immoral? Look at today's world and ask yourself whether anyone is able to produce a compelling proof of its morality or immorality, a proof starting from universally accepted premises and showing by purely logical means that one of the two systems is the just one. Now how about the alleged intuition, the contemplation of the good, the inner evidence?

What is moral: being faithful to one's wife and closing one's mind to love, or seeing something divine in love and being open to it? Was Goethe immoral because he loved many women? Think of what theologians, moralists, poets, playwrights, politicians and founders of religions have said about these things, in all sincerity and after thoroughly examining their conscience, and note the enormous differences between their convictions – do you still want to appeal to moral intuition?

The differences are even greater if we compare different cultures with one another. One example may stand here for many. It comes from Darwin, who tells the story of an African 'savage' feeling terrible pangs of conscience because he had neglected to take revenge on a neighbouring tribe for some act of magic.

A missionary had impressed upon him that it is a great sin to murder a man, and the savage did not dare to carry out the act of vengeance. But the consciousness of his neglected 'duty' oppressed him so much that he went about disturbed and upset, rejected food and drink, and could enjoy nothing. In short, he showed all the signs of an 'evil conscience'. Finally he could bear it no longer, stole away secretly, slew a member of the other clan and returned light of heart: he had performed his duty and pacified his conscience by means of murder.[7]

Who will claim that the feelings of the 'savage' were not genuine pangs of conscience?

But the most telling case in point from our own time is the confrontation between Nietzsche's ethic and the Christian ethic. Nietzsche calls bad everything Christianity regards as good: humility, suffering, peacefulness; for him, evil, hardness, dominance are the real good, and they are only reputed to be bad because of the resentment of the weak and backward – the revolt of the slaves in morals. Can there be any doubt that Nietzsche really meant it? In criticizing this doctrine, are we to appeal to the fact that proclaiming a 'master morality' and dividing men into masters and slaves is contrary to the principles of justice? We would not gain anything by this; for Nietzsche tells us that justice is an invention of the poor and disadvantaged, who avenge themselves on their masters by poisoning their conscience. Clearly, we have here one ideal set against another, and one conviction set against another.

Impressed by such examples, one begins to doubt whether there is an intuition that teaches us what is right and wrong. To be sure, experiences of conviction play an important part in one's acquisition of a moral view. But such experiences cannot be the *criterion* for the validity of a morality, for they differ after all from one person to another. A Christian and a Nietzschean could both declare that they are thoroughly convinced of the correctness of their view. Actually, anyone who defends a moral position different from the usual one with honest zeal could be used as an example.

[[Defenders of the theory of self-evidence claim of course that in this case the subject did not experience the true, genuine evidence, so that his conviction was not backed by any genuine evidence. But this claim gets entangled in an insoluble contradiction. For either genuine evidence is experienced as something decisive, different from false, merely subjective conviction, so that they cannot be confused at all with one another; and in that case illusions of evidence do not occur at all, which denies the state of affairs the whole theory was invented to explain. Or else there is no demonstrable difference between the two experiences or states of consciousness. This means that it can be decided only indirectly, i.e. by subsequent investigation, whether an ethical conviction was or was not backed by genuine evidence, and this is to admit that the criterion for the correctness of a conviction is not to be sought at all in the experience of evidence, but in something else, namely in the criteria that had to be [used] in that subsequent investigation. But this invalidates the claim that self-evidence is the bedrock of moral knowledge. Thus both alternatives lead to contradiction with the presuppositions of other theories, and the result is that the actual distinction between conviction with and without evidence was merely an artificial construct, devised to prop up the claim that any correct moral position announces itself to us by a special unmistakeable experience of evidence.

This explains why we cannot today accept an appeal to self-evidence as the supreme judge. To repeat, there are many incompatible, mutually contradictory ethical systems:

(1) Epicurean ethics,
(2) Stoic ethics,
(3) Christian ethics,
(4) Nietzsche's ethics.

And this of course raises the question: can science decide between them? Do its results make it possible to find out which morality is 'correct'?]]

If in the history of science or philosophy a problem stubbornly resists all attempts to tackle it, there will sooner or later arise the suspicion that the correct way to put the question has not yet been found. We know today that a large class of philosophical problems owe their existence to an incorrect formulation. And this brings us to a more essential point: When it is asked which of the various ethical systems is correct, or whether it is possible to demonstrate objectively the correctness of an ethical system, is this question put correctly in this form? For ethics usually assumes the form of precepts, of moral prescriptions. But can precepts or prescriptions be true at all? What methods do we have for deciding such a question? I believe that we first need to reflect on what we *mean* by a norm? We must examine the *meaning* of a norm, or the *meaning* of a value judgment.

We say: 'That is a really good deed' or 'That is a noble motive'. This sounds like a description, and yet it merely expresses our approval. Here, as so often, we are misled by language. When someone declares that for him life is meaningless and has no value, that he approves of anything that would lead to the extinction of the human race, there is absolutely no way to refute him. We can indeed try to make him see the bright side of life so as to change his attitude; but that is something completely different from refutation. What we do in such a case is to lead him away from his gloomy thoughts. Suppose you know that a man is contemplating suicide and you try to talk him out of it and get him to change his mind. What can you do? Can you *refute* him? Can you *prove* to him scientifically that his pessimism is unfounded, that life is good and worth living? Would you for example argue that he is *value-blind* and lacks the evidence to back up his judgment? You will do nothing of the sort. You can help him, talk to him in a friendly manner. You can encourage him and plant fresh hope in his mind by showing him a way out where everything seems blocked to him, or you can *console* him by showing him purely human sympathy, compassion and friendliness – in short, you can counteract his despair and try to change his *will*. But the essential thing is that you *do* something – and do not try to convert him theoretically.

This brings us to the decisive point: ethics is a matter of the will, not of the understanding. This is why ethical sentences have nothing to do with knowledge and error nor with 'true' and 'false'.

The approval about which philosophers of value talk so much does not consist in taking something to be true, in that passive contemplative state we are in when we abandon ourselves to looking at a geometrical figure and finding a theorem to be true; there is something active in approval: it is affirming something, taking a position, supporting it and taking it upon oneself, making a commitment; and what speaks through it is the voice of the will rather than that of knowledge. The fact that our language calls such different acts by the same name – assent and dissent, affirmation and denial, acceptance and rejection – has helped to create much of the age-old confusion in this field.

If we want to investigate this question scientifically, we must proceed step by step and observe how, e.g., the word 'shall' is used in our language. Here I would like to compile a few perfectly simple remarks which will perhaps elucidate the meaning of that word. In the use of our language there is a close connection between the words 'yes' and 'no' or 'true' and 'false', which comes out, e.g., in the fact that we can say:

> It is true that it is raining = it is raining.
> It is false that it is raining = it is not raining.

I.e., it is characteristic of the grammar of the words 'true' and 'false' that (at least in a large class of cases) the affirmation or negation of a statement can be replaced by the declaration that the statement is true or false. For example: he did not kill = it is false that he killed. We now introduce the word 'shall' into this context; and we notice at once a most striking phenomenon: the negation can no longer be expressed by the term 'false'. The biblical commandment reads: 'Thou shalt not kill'; but it would not occur to anyone to say: 'It is false that thou shalt kill' – a hint that negation here takes on a different meaning and loses its connection with 'true' and 'false'. It is also significant that some languages like Latin and Greek use one word for the negation of a statement and a different word for the negation of a norm.

Another fact that belongs here is the following: A child learns what he is supposed to do when adults tell him that he *should* be obedient. But he will hardly ask: is that true? He can for example ask: 'Should I really do that?', i.e. 'Do you mean it seriously?', but not: 'Is that prescription true?' I would say that the child is much clearer about the meaning of the word 'shall' than are those philosophers who use much ingenuity to philosophize truth and evidence into the 'shall' and thus distort and twist the original meaning of the word. Norms have after all

an unmistakeable resemblance to commands and wishes and are often verbally indistinguishable from them; and nobody will want to assert that imperative or optative sentences are true or false. But this is not to say that norms are only a special class of rules in the sense in which, e.g., even numbers are a subclass of natural numbers; I am only pointing to a certain similarity[8] [[in the grammar of 'norm' and 'rule'. It will hardly do to say that rules of grammar or definitions are on a very different level from that of ethical norms. The best way to clarify this difference would be to describe the role or function norms have in the life of a community. Thus one will have to say that in growing up a child learns the ethical norms of the community, connects them with reward and punishment, and is taught to feel respect and awe before certain actions and disapproval and revulsion or shame before others.
[...][9]

That a morality cannot be justified becomes obvious if we ask ourselves what objections can be raised against a particular morality. Let us take Nietzsche's ethics for example.

First, it could be said that, concretely understood, Nietzsche's ideals are clear and great and [...]. Let us imagine a community in which the following has been established as a norm: 'Whatever members of the master race do is permitted. The rest shall serve the members of the master race.' It is clear that there can be no administration of justice in such a community. What Nietzsche had in mind was the following idea: whoever feels in his heart that it is his calling to become a master and has the courage to overcome the slave morality will be a member of the master race. This would be the criterion of being a master. If based on this criterion we follow up the purely logical consequences of the above norm, we see that members of the master race could establish a slave morality [...]. But how are they to deal with conflicts between themselves? Suppose they create a law, a statute book for masters, to settle such conflicts. There will therefore exist two legal systems side by side: one for slaves and one for masters. But what if a master is insulted or robbed, etc. by a slave? Suppose it is established who perpetrated the act. But the perpetrator, whoever he may be, can avoid punishment under the slave laws by claiming that he is a master, and he could prove this claim precisely by the fact that he had the courage to overcome the slave morality and shout at the master. Hence the criterion Nietzsche [had in mind] would make the application of the slave morality illusory; i.e. for purely logical reasons, the criterion he [had in mind] is incompatible with the assumption that there are two different moralities.

A believer in Nietzsche's ethics could avoid this difficulty by giving a different criterion distinguishing between masters and slaves, one making it impossible for a person to declare himself to be a master: [...] precise [...] characteristic marks: self-determination, outstanding achievements, etc. In that case such a legal system would be logically possible.

Secondly, it could be said that the rule of the will to power, as Nietzsche envisioned it, would not have the intended effect at all, for Nietzsche had in mind a particular state of culture, one he found realized above all in the sixth century before Christ. It could be said that the sixth century he so greatly admired owed its achievements to very different circumstances from the naked will to power. It is more than likely that the flowering of culture at the time was based on trade, colonization and the beginnings of progress rather than on the will to power, which was frittered away on countless local wars that lacked any cultural value, ruined the land, and even contributed to the downfall of the great public figures.

Such a [consequence] would oblige the believer in a master morality to make a new decision: he would have to make it clear to himself what he really wanted: unbridled will to power or great cultural achievements. By pointing out such consequences to him, we could at best make him see that the drive for power did not have the effect he expected it to have, and it would now be entirely up to him whether to stick faithfully to the ideal of the will to power or accept some other morality under whose rule there is a better chance of realizing the ideals of that century. But in the end, this is again a matter of decision.

Thirdly, it could be said that the theory of eternal recurrence, to which Nietzsche attached such importance in his later years and which he regarded as the metaphysical underpinning of his ethics, is really meaningless. For if all things and states of affairs in the world recur exactly as they were, one could just as well say that everything in the world is just what it is and that time is self-contained. But Nietzsche could of course have maintained his ethics of domination without burdening it with those metaphysical speculations.

It will be seen that such criticism touches only the *expression*, not the *intention*. Whether I sympathize with pride, toughness, militancy or with Christian virtues is not decided in this way. Nietzsche uses every means in his language to talk us into accepting his theory of values, to *seduce* us, but he does not adduce a single scientific argument. You can tell right away that you cannot make a dent in his attitude to values by scientific considerations and arguments. In the final analysis it always comes down to the question: Do you will the world to be like this? Or do you will it not to be like this?

As far as I can see, a system of norms can be criticized only as follows: by trying to prove that the norms are logically incompatible or have been expressed unclearly, that the goal has to be inferred from the application, or that the causal connection has been misjudged so that the results desired by the legislator fail to materialize. To criticize it by asking whether the norm is *correct* or whether it is *worth* establishing is merely playing with words.

Take a simple example. I can ask whether a move in chess is correct, i.e. according to the rules: this can be decided definitely by comparison with the rules of the game. But I cannot ask whether the rules of chess should be as they are: such a question violates the very meaning of the words. And yet ethical investigations consist in large part of such questions.

But, you will reply, the norms of ethics are not rules of a game: they are eternally valid imperatives, like the laws of logic. Well, let us look at the situation in this field.

To begin with, logical norms are certainly not eternally valid imperatives. For it should be clear today that, strictly speaking, there is not just *one* system of logic, but different logical systems differing from one another in various details: there are systems with and without the theory of types, with and without the axiom of reducibility, with and without the law of excluded middle, etc. The idea of *one* logic belongs to a backward state of science and is today obsolete.

I can ask whether a given proof is correct; the word 'correct' here means: according to the rules of logic. Logic gives directions, rules, norms, which inferences and proofs are supposed to follow.

The question whether a proof is correct has a clear meaning. But now imagine someone raising the question: Are the rules of logic themselves correct? Or should these rules perhaps be formulated differently? Such a question leaves us in a quandary. I would not say that it was meaningless – the word 'meaningless' should be used sparingly – all I am saying is that in this form it is impossible to see clearly what the questioner has in mind. But we could, if we like, give the question a different meaning.

Imagine for example someone setting up a different system of rules, as is actually being done today. The question whether we should accept this newly created system of rules and use it to make inferences will depend on what we require of a system of logical rules and what the new system enables us to do. Suppose, however, it turns out on closer inspection that the new system does indeed have certain advantages over the old one, then we could also say of it that it is a better logical system. We have then evaluated the two logical systems – but this evaluation is again possible only on the basis of a previous agreement specifying which qualities a logical system *should* possess, and this convention is *again arbitrary*. We will not succeed in producing a justification if only because it would have to come to an end, i.e. lead in the end to something that could no longer be justified. Thus we never reach anything final. The last thing we reach is only a stipulation.

I would put it like this: in logic we do not want a justification because no justification can satisfy us. What we can give is, in the end, only a description of logic. Logic is autonomous. Whatever looks like a justification is thus already [...] and will not quite satisfy us. We never ask for a 'why'.

An ethicist can proclaim his doctrine and try to promote it [...] by persuasion or by living by it. But there is one thing he cannot do: he cannot justify his ethics. Schopenhauer says: 'it is easy to preach, but difficult to found, morality'. I say: it is difficult to preach, but impossible to found, morality.[10] Ethics, like religion, is something you can only *profess*.[11]

A final word about norms: by saying that norms can be freely *chosen*, by not regarding them as eternal verities, I seem to have abolished their rank and demoted them. I do not know what to reply to this reproach. The objection rests on an estimate of the value of ethical norms, and any discussion of it would go beyond the bounds of science. But if in concluding this lecture you permit me to step outside the bounds of science and profess my values, I would say: No, this conception of norms does not devalue them in my eyes. On the contrary, the sharp division between the voluntary and the cognitive has the effect of *sublimating* ethics – it makes it clear that ethics consists of imperatives which we *profess* and which we therefore accept in a very different way from the way we accept external laws.

Science neither destroys nor protects norms, but tries to make us see clearly that these peculiar entities cannot be justified; it raises our awareness by making us see that there are many different possible ethics, and while it gives us *freedom*, it also gives us *responsibility*. For, let me make one thing clear, anyone who would conclude from the preceding considerations that ethics is basically arbitrary, just like the rules of a game, would be making a big mistake. Professing a morality is a *deep process* – I know of no other word I could use to describe the feeling we have that the core of our personality expresses or manifests itself in this choice.

People who follow an ethical system without question, without giving it a second thought, whose lives run along smoothly and in a straight line – how much easier it is for them than it is for us! They have no inkling of the enormous variety of possible decisions and the dangers implicit in them. All they have to do is to follow ethical imperatives, and this is where their merit lies. But for someone whose consciousness has been set free by knowledge and who sees the excitements of life and its ideals and has to *choose* between them – for such a person ethics acquires a new meaning and new seriousness. The tissue of half-truths woven by human imagination around moral laws vanishes before his eyes; he stops asking for the truth of morality and starts choosing and deciding. And does this not give ethics a new dignity alien to the old conception?]]

NOTES

[1] Arthur Schopenhauer, *On the Basis of Morality*, translated by E.F.J. Payne, Indianapolis: Bobbs-Merrill, 1965, p. 40–41. [TR.]

[2] I am not saying that scientists are anti- or a-metaphysical. This is a matter of personal attitude. But the general tendency is undoubtedly for science to detach itself from metaphysics and make itself independent – if only out of a kind of instinct for self-preservation, for science does not want to be drawn into the anarchy and struggle between metaphysical doctrines.

[3] Max Scheler, *Der Formalismus in der Ethik und die materiale Wertethik*, Halle: Niemeyer, 3rd ed. 1927, p. 20 [EDD.]

The work is available in English as *Formalism and Non-Formal Ethics of Values*, translated by Manfred S. Frings and Roger L. Funk, Evanston: Northwestern University Press, 1973. There the quoted passage reads as follows: '... *clearly feelable* non-formal values of their own kind. Of course there is nothing subject to definition here, as is the case with all basic value-phenomena. All that can be requested is that one attend to seeing precisely what is immediately experienced in feeling good and evil' (p. 25). [TR.]

[4] *Ibid.* [EDD.]

[5] Oskar Kraus, *Die Werttheorien*, Berlin/Wien/Leipzig: Rohrer, 1937, p. 172 [EDD.]

[6] Cf. K[arl] Menger, *Moral, Wille und Weltgestaltung*. [E.T. in the present series, *Morality, Decision, and Social Organization*, Dordrecht/Boston, 1974 – EDD.]

[7] Moritz Schlick, *Problems of Ethics*, translated by David Rynin, New York: Prentice-Hall, 1939, p. 91–92. [TR.]

[8] The MS contains the following variant: '(... but I am only pointing out a certain similarity) with imperatives. Logical norms are not orders issued by someone, and it seems forced and unnatural to conceive of them as imperatives. But there is another difference: There is a power behind every command which gives it force. Without this power, a command would not be a command. But a norm does not presuppose a power; we are free to accept it or reject it./But this is not to say that every norm is only a special class of rules, in the sense in which all odd numbers are a subclass of numbers; I am only pointing out certain *similarities* in the grammars of norms and rules.)' [EDD.]

[9] The MS continues (in so far as it is legible): 'To make it easier to reach a decision, let us look at [...] in another science. I would call your attention at this point to a remarkable similarity between ethics, law, logic, geometry and grammar. Law can be understood in two ways.' After this, several pages (probably five) are missing in the MS. [EDD.]

[10] It is not certain whether this paragraph and the following sentence belong here. They seem to belong to a development of some ideas of Wittgenstein's. The quotation from Schopenhauer is from *On the Will in Nature* in: Arthur Schopenhauer, *On the Fourfold Root of the Principle of Sufficient Reason* and *On the Will in Nature*, translated by Mme Karl Hildebrand, London: George Bell, 1889, p. 372. For the original formulation of this response by Wittgenstein see Friedrich Waismann, *Wittgenstein and the Vienna Circle*, edited by Brian McGuinness and translated by Joachim Schulte and Brian McGuinness, Oxford: Blackwell, 1979, p. 118 [EDD.]

[11] We can accept an ethic and profess it; but if someone rejects it, we cannot *prove* that he is wrong; or better; it makes no sense here to speak of proofs. [See Waismann op. cit. loc. cit. for Wittgenstein's idea that in ethics we must speak in the first person (='profess') – EDD.]

FRIEDRICH WAISMANN

WILL AND MOTIVE

TABLE OF CONTENTS

Editors' Foreword	55
1. Freedom of the Will	55
2. The Concept of the Will	57
3. Types of Willing	62
4. Am I Master in My Own House?	67
5. [Characteristic Marks of Willing]	70
6. Willing and Wishing	74
7. Willing and Knowing	77
8. On the Indeterminacy of Willing	84
9. The Divided Will	88
10. Will and Ego	89
11. The Ambivalence of the Will	94
12. How Does One Come to a Decision?	97
13. Motivation	107
14. How Does One Come upon a Motive?	111
15. Having a Motive and Being Motivated	118
16. Fathoming a Motive	119
17. The Existence of Motives	125
18. Emotion and Action	129
19. Motive as Interpretation	131
Notes	137

FRIEDRICH WAISMANN

WILL AND MOTIVE

FOREWORD

The material printed under this title (which was chosen by J.S. for the German edition) exists both as a typescript and as a carbon copy of the same. The latter contains a few corrections by Waismann which are not present in the top copy and was therefore used for the German edition from which the translation was made. Editorial additions and interventions are enclosed in square brackets. All notes have been supplied by the editors [EDD.] or the translator [TR.]. – *The Editors*

1. FREEDOM OF THE WILL

At some level of our being we experience choice; that is, we all know situations in which we say we make a decision and hence choose between different possibilities. Now if I were to say that the act of choosing is the final deciding factor, I would be a believer in free will. But I do not believe in such a mystical act, an act that puts an end to vacillation by, as it were, tipping the scales. For we cannot rule out the possibility that there is something else behind the choice and the conflict of motives: innate dispositions, enduring traces of previous decisions, momentary influences arising from our personal lives, influences which can perhaps be felt but are difficult to grasp, interests or inclinations we are not clear about, or would not like to admit openly and without reservations. The decision I make may not therefore be the result of a simple act of will – would it not perhaps be more correct to call 'will' the whole process of decision-making? But if this is so, what could be meant by freedom of the will? *What* is supposed to be free in this whole complex process? Is it the will? But the will is precisely the whole variety of processes, inner as well as outer, linked with the antecedents from which the decision arises. In short, although experience shows that we choose between possibilities, this does not prove in any way that this choice is something final and uncaused, that it is not in turn conditioned by many other factors – perhaps even by processes occurring at a deeper level.

Should I therefore say that there is no freedom of the will? That everything I do is determined? Here I am faced with a new difficulty: what is meant by 'determined'? That is, how can I decided whether my choice or decision is determined by everything I have experienced up to now, or whether it is not yet determined by it? Is there some characteristic by which I can distinguish one of these possibilities from the other? What is the criterion of being determined? You probably have the following idea: I cannot, strictly speaking, decide between different possibilities; my idea that the choice is in my hands is an illusion, for at bottom the way I am going to act has already been decided by my whole previous life; my way has, as it were, been marked out before me, but as I do not know this yet, I therefore toy with the idea that I could go different ways. But all these considerations would have a clear meaning only if there were some criterion for verifying whether I really had no choice and had to act precisely this way and no other. Now what would such a criterion consist in? In the inanimate world we do not speak of acts of will, but assume that observable events are causally determined. Now the criterion of causality is *predictability*: if my knowledge, including that of the laws of nature, enables me to predict what would happen under certain circumstances, then this event is said to be *causally determined*. Now I am perfectly willing to admit that a person's actions or decisions would be determined if it were possible to predict his behaviour; that is, if we knew laws on the basis of which we could calculate what he would do before he did it with the same certainty with which we can make predictions in physics. Thus if we could say that under such and such circumstances this person will act in such and such a way and under those circumstances in that other way, irrespective of any struggles or conflicts going on inside him (for the formula has already taken all this into account) – then and only then would it be shown that he really had no choice and that the belief in freedom of the will was an illusion. Now in actual fact we cannot predict a person's behaviour, or only with some degree of uncertainty. In thinking of a person we know very well, such as a close relative, we can indeed say: 'I am sure that under such and such circumstances he will do this but not that', but it is not hard to imagine situations in which we do not quite know what a person will do, no matter how well we know him, and where we would say instead: I am fairly sure that he will *not* do this; but he will perhaps do *that* or *that* or *that*; I think it improbable, but not impossible, that he will also decide on *that*. In other words, our actual situation is such that in the most favourable case in which we know a person very well we can picture the possibilities open to him and exclude certain other possibilities, but we can never reduce the realm of possibilities to a single one, to the way he will actually go, and there is no indication that this will change as our knowledge increases; for our fundamental picture of a person at the moment of decision seems to remain unchanged: before him is an open cone of possibilities, some more probable, others less so, but all of it uncertain and fluid, without sharp boundaries. But if we cannot predict a person's actions, then any talk about their being absolutely determined will make no sense; indeed, how am I supposed to *distinguish* the case in which a person's actions are determined from the case in which they are not, if I can no longer use the only means of making such a distinction, namely *predictability*?

I cannot therefore say either that human beings are free or that they are unfree. Both views are false, both pictures are at variance with reality. Free – unfree: these are predicates of actions (I act freely if I am not constrained) rather than of the will. Freedom enters, as it were, into a chemical bond with the will.

Incidentally, the perspective in which we see things changes: the moment I am faced with an important practical decision I will be imbued with the feeling that this decision comes entirely from inside me, that the choice is, as it were, entirely in my hands. When I look at the same decision again from a temporal distance, free of all the tendencies that governed me at the time, and when I view my own past with a knowing and searching eye and call to mind what I was and what forces and influences impinged on me, I will perhaps be inclined to say that my decision expressed the whole direction of my life at the time and could not very well have turned out otherwise. To have acted differently I would have had to be a different person.

2. THE CONCEPT OF THE WILL

Is there an experience of willing? And how does it differ from an experience of wishing? Is it perhaps like this: wishing may at first be weak, get stronger, and keep increasing in intensity, until its intensity exceeds a certain limit and it turns into willing? That is, is the will only an extraordinarily intense wish? Obviously not; I can wish for something passionately, long for it with every fibre of my being, burn with longing for it – and yet my longing does not turn into willing. What is the difference?

Let us take as simple a case as possible. I am lying in bed in the morning and would like to get up. I see that it is already quite late, I have much to do, and I should be getting up. I take a run-up, as it were, and – remain lying. The room is cold and uncomfortable, I feel tired – in short, I do not get up. And yet I sincerely wish to put an end to this to-ing and fro-ing and get out of bed; I even have a very strong wish to do so, and I am cross with myself for still being in bed; perhaps I have the wish to be the kind of person who could jump out of bed without any difficulty. But all this remains in the realm of wishing; I have not yet willed. I have willed only *when I actually get up*. As long as I only considered getting up, imagined it, or wished for it, what I was doing was no more than wishing. What had to be added to turn it into willing was the act.

Many philosophers have worried over the question how the will manages to set the parts of the body in motion. They have assumed that the will is something mental and a movement something physical. How, they ask, can a process in the mental realm, which is non-spatial, e.g. a volition, have an effect in the physical realm, e.g. generate a movement? Looked at in this way, the matter seems miraculous. I now want to raise my arm; that is an experience in me, a volition; and now, lo and behold, my arm actually rises, as if raised by an invisible force. I really ought to be beside myself with astonishment, like a witness to a miracle, wide-eyed, unable to understand it in the least. It is as if the invisible idea in me had suddenly materialized as if by magic. What! the mere idea of raising the arm causes the arm to rise – what a magical kingdom we live in! But stop! As long as I do not raise my arm, I have not even willed: I may perhaps have wished it, thought of it, imagined it, etc. I have willed only if I actually execute the movement; so *the will is the act.*

It is not as if there were on the one hand a clearly recognizable experience, the volition to do something, and on the other hand the act of doing it, and as if the first produced the second; for the criterion that distinguishes willing something from wishing to do it is, precisely, the doing of it.

This makes it look as if willing was always preceded by wishing or intending, and as if we spoke of willing where the wishing or intending was realized. But is that so? Is willing always preceded by a thought, an intention, a wish? Suppose someone comes up to me and offers me something for sale; I decide to buy it without thinking; in that case I willed; but my willing was not preceded by either an intention or a wish. When, then, do we speak of willing?

In daily life we do thousands of things, as we say, mechanically, without any question of willing. At the table I reach for knife and fork, pour myself some water, or pass someone the sugar bowl, without any volitions coming into play. All this is done mechanically, has been practised thousands of times and has become conventional, and it is now almost as much of a reflex action as when I raise my hand to brush away a fly that has landed on my nose. In most cases we do not will but simply *do* something. Willing comes in only when there is some *blockage* in this smooth flow. Suppose I reach for the fork but cannot pick it up. This catches my attention; I try once more, look at it carefully from all sides and this time I try to pick it up by force. I feel some resistance, experience muscular tension, and overcome the resistance. (Some practical joker may have played a trick on me by fastening a strong magnet under the table to hold the knife in place.) In this case it will be said that my will is directed towards grasping the knife, that I do not, as I usually do, pick it up automatically. If there were no resistance, if everything I did went smoothly and without any effort, there would be no experience of willing. The first lesson to be learned from this is that willing always goes with some resistance and is directed towards overcoming that resistance.

But did I not just say that I can decide to buy an object I am offered without giving it a thought, and that I shall have willed in that case? Where is the resistance here? Well, for a moment the question whether to accept or decline the offer will have arisen in my mind; even if this hesitation lasted only a brief moment, at least there was a conflict between two ideas in which one idea proved stronger than the other and suppressed it. There was therefore an inner inhibition the overcoming of which made this process one of willing. If there was no conflict whatsoever between ideas, if I did not hesitate, however briefly, between acceptance and rejection, my 'yes' would have been, not a decision, but just a mechanical reaction like that of a conductor issuing a ticket in response to a passenger's request. This goes very well with the following observation: If I say 'I will eat an apple', the element of willing has almost entirely disappeared; it almost says: 'I am on the point of eating an apple' or 'I am about to eat the apple'; this is why willing has become an auxiliary verb in English. We might say that the will has degenerated into the future tense. On the other hand, it may take an enormous effort of will to do nothing – to remain calm in the face of a serious insult, or as in previous wars, for soldiers to stand in the front line and calmly accept being shot at. We admire such strength of will just because we can see signs of powerful forces being overcome.

But what if an unscrupulous businessman is offered a good buy and grabs it without hesitation? Must we first examine carefully whether there was a 'struggle' between two ideas in his mind before we can say that he willed it? I believe that two ideas cross each other in this case: the outward situation of the purchase, as it appears for example to a lawyer, and the inner psychological situation, which alone is at issue here. In the legal sense it can be said that the businessman is responsible for the purchase, that it can be ascribed to him or to his 'will'. But if we speak of the 'will' in this case, it only means that he was free to choose between saying 'yes' and 'no' and not that such a decision actually took place in him. To 'ascribe the purchase to his will' only means holding him responsible before the law and recognizing him as a legal subject. An under-age child for example (or a slave in Ancient Rome) has 'no will' before the law. This is not, of course, a psychological statement. Thus in the legal sense the purchaser uses his will whether or not he suppresses an inhibition in making the purchase. However, in the psychological sense we can talk of a will only where there is an outer or inner inhibition being overcome. So if the will is the act, the act is by no means the will.

Is the case of the will, then, like that of courage? Is a person courageous only if he somehow feels fear and overcomes it? Or can a person be courageous without even knowing what fear is? Or is only this true courage?

Maybe people like Napoleon who appear very strong-willed to the outside world show no such remarkable efforts when seen from the inside, because they pass from intention to action easily and without much inner inhibition. Perhaps the natural course of events inside them is so uninhibited that there is no real inner conflict and no real feeling of overcoming or effort, and hence no real experience of willing. On the other hand, people who are said to be 'weak-willed' by nature may often have to fight it out within themselves before they are able to make up their minds, and they may have to make heroic efforts to force themselves to make even the smallest decision. Perhaps no other group of people makes more desperate efforts than hopeless failures like alcoholics, drug addicts and 'weak-willed' dreamers who constantly run up against their own nature and constantly succumb to it.

Of course, it may also be that the great historical figures were really strong-willed and therefore able to cope with any resistance. How are we actually to decide between these two possibilities? What are we to make of the statement, 'The only strong-willed people are neurotics, not heroes'?

But let us return to ordinary people.

Suppose there is an enormous resistance to be overcome. Can I for instance direct my will towards becoming a conqueror like Alexander the Great, or a playwright like Shakespeare? Or can I direct my will towards overturning the present government? Such goals are obviously unattainable to my will; I can perhaps indulge such wishes in the imagination or dream about such things; but can I seriously direct my will towards them? If so, where should it start, and what should it do to attain one of those goals? It could be said that such a goal is entirely outside the scope of my will. So there are situations in which there is no willing, not because of lack of resistance, but because of overwhelming resistance.

This teaches us a second lesson: what we call 'will' lies between two limits: between a certain very small resistance below which we do *not yet* speak of willing and a very great resistance beyond which we *no longer* speak of willing.

But are these limits fixed? No; it is rather like this: the idea of willing evaporates at the two ends without a sharp boundary; however, the limits of the will can be illustrated by examples. I am taking off my collar, but the button 'sticks' and refuses to budge; thinking about other things, I tug at it mechanically for a while, but it still refuses to budge; finally I give it my full attention and try hard to open the collar. Can we draw a sharp boundary between mechanical and voluntary action? Is the decisive moment perhaps when the problem catches my attention? But my attention comes and goes in waves, turns perhaps momentarily to the collar and bounces off again or continues along the former track. It is just that mechanical and voluntary behaviour are connected by a series of links, without a sharp break.

Can I will something completely outside my power? I cannot will the sun to rise right now, or the ocean to recede, or my heart to beat faster. Am I to conclude from this that I can will only what is in my power? That would not be quite right. Suppose I believe in magic or ascribe unusual powers to myself. Then I could will my heart to beat faster. I could even stare at the vase at the other end of the room and tell myself: 'I want it to fall over by a mere effort of my will.' Am I to say that in this case I did not really will but merely wish? This would not be correct, I think. I did in fact do something like willing. But what, then, did my willing consist in? Well, in my looking steadily at the vase, in a feeling of strain pervading my body, in a tightening of my muscles, in my making a determined face, in my focusing insistently, stubbornly on the object, and in my saying to myself under my breath: 'You've got to fall over now.' Something like an effort of will is thus possible even in this case. Thus instead of saying that I can will only what is in my power, it is more correct to say: I can will only what *I believe* to be in my power. Since what I assume I am capable of doing varies with my mood, world view, etc., there must also be some indeterminacy in the concept of willing. Suppose I say: 'I will lift a ton.' I try it and fail; did I will it? If I knew in advance that it was hopeless, I only acted as if I willed it; whereas if I set about it with full confidence, I did will it; but what if I was uncertain? Did I or did I not will in that case? Can I say: 'I will to be loved'? This question is difficult to answer. I can perhaps believe that it is in my power to be loved; I can even do a few things along that line – but is that all? Or are there perhaps different types of willing? Does language perhaps combine several essentially different processes into a kind of whole? Does the will to be loved not differ *entirely in kind* from the will to stretch out my arm? Can willing and wishing be mixed?

3. TYPES OF WILLING

Let us consider a few examples: I *will* remember something; I *will* deal with a mathematical problem; I *will* not think of it; I *will* revenge myself for an insult; I *will* start a new life and become a different person; I *will* submit to my fate; I *will* yield to evil; I *will* believe in God; I *will* not live; I *will* return to the past; I *will* not *will*.

'I will remember': I make an effort to become conscious of a forgotten name or event; perhaps I direct my thoughts along a certain path and scour the whole area around it in the expectation that the missing item will turn up. Willing here consists in making an effort, in concentrating. But since I do not know for sure that it is in my power to remember it, my willing becomes somewhat dubious. We therefore say: 'I am *trying* to remember.'

'I will deal with a mathematical problem': I set aside some time, maybe cancel a visit, pick out some books, get paper and pencil ready, focus my attention on the problem and try to concentrate. In this case, willing consists not only in concentrating one's attention, but also in taking some preparatory steps: clearing away obstacles, arranging to be free, etc.; in short, the willing here is not a single act, but a whole series of processes which may occur at different times; it cannot be said of any of these processes that it would amount to willing by itself, without any of the others present: for I would not say for example of a suddenly awakened interest in a mathematical problem that it was the will to deal with it; and if I now reserve some time, lay out pencil and paper, but give no further thought to it, I might say that I *thought* of dealing with the problem, but not that I actually willed it. But suppose I was disturbed during my preparations so that I did not get to deal with the problem – what then? Do I myself know with certainty whether I willed or merely let my thoughts drift in that direction? That would seem to depend on the whole background of my life, e.g. on whether I am used to dealing with mathematical problems, or whether I had similar experiences in my past life showing that this time too it was probably only a false start.

This example shows two things: first, the will need not be something simple like an 'impulse'; secondly, I myself cannot say with absolute certainty whether or not I willed. The complexity of the situation becomes even clearer if I change the example a little and say: 'I will study mathematics'; this presupposes many actions, preparations, interests and perhaps even changes in my way of life; or better, the will includes all that.

'I will not think of it': I try to avoid everything I associate with it and chase away the thought of it when it occurs to me, perhaps I try to take my mind off it, or deal with something else, or throw myself into work; I avoid any opportunity to speak of it, and even inwardly I dodge any such occasion. I am conscious of a negative attitude to the whole region where the thing I do not want to think of comes from; the whole region is, as it were, taboo. But in a weakened state, when I am tired or about to fall asleep or overflowing with emotion, the taboo thought can catch me unawares; I put up a fight, but it is stronger and I am weaker, and I cannot shake it off; and thus there can be a struggle between the emotional force that tries to make me conscious of the thought and the defence mechanism I have set up against it in my consciousness. This shows again how far removed the will is from a simple act.

'I will revenge myself for an insult': If I react to an insult emotionally or, as we say, 'impulsively', we shall hardly say that I *willed* to revenge myself. Rather, the will to revenge myself presupposes that I nurtured feelings of revenge, entertained the idea of revenge, considered various possibilities, and finally chose one of them. It thus presupposes an inner turmoil, a certain imaginative effort, some intellectual work – examining the various means and choosing a suitable one – and one more thing: the decision to translate the approved plan into action. Now is it not this final decision which actually constitutes the will? But we would not speak of a decision or the will to revenge if it had not been preceded by certain passionate feelings, certain images of the revenge having been taken, as well as by some planning. For the whole thing is a more or less long process, the taking-shape of a plan plus the final go-ahead; but what if I shy back at the last moment and do not go through with it? In that case I have only toyed with the idea of revenge, wished to revenge myself, but not willed it. The decision consists in the act: but the will to revenge myself is not just this act, but the act together with everything that precedes it and produces it as a result. The will comprises all these processes and is not identical with any one of them.

'I will start a new life and become a different person': This is very complicated. It involves dissatisfaction with my previous life, remorse, insight – or supposed insight – into my path to salvation, assessment of my own personality and an estimate of my own power, knowledge of other people's fate, preparations, deliberations, and finally a firm decision, a change in my way of life or attitude to life, the beginnings of action, and steady progress along what I know to be the right path: all this is gathered together in what is here called 'will'. One more thing: only if I let myself be led away from that path, but feel remorse and return to it, do I really know that I willed and did not just have a pious intention. In fact, the depth and duration of remorse can be regarded as the measure of the strength of the will to change one's ways. This shows that one can really know only in the light of future experience whether one has willed, and one cannot directly perceive the presence of such a will by, as it were, being immersed in one's own inner self. No introspection, no investigation of my own inner self, however unsparing, can convince me that the will to change my life is really alive in me. For this knowledge requires data that are not yet available to me at the moment I examine myself and will become accessible to me only in the future. Here the phrase 'Know thyself' is really nothing but a pretty figure of speech.

'I will submit to my fate': The striking thing here is that 'submitting' or 'accepting' is virtually the opposite of willing. In fact, this sentence can mean very different things: it can express a mood of resignation and renunciation: 'I am tired, I have made so many efforts, all to no avail, I see no other way, I give up'; but it can also express an insight: 'I have finally realized that my lot, hard as it is, is probably the best thing destiny has to offer me and that it would be sinful to rise up in revolt against it'; or else it can express a resolution: 'I will refrain from grumbling about destiny, I will stop trying to rise up against it, I will try to accept this life and make the best of it.' And all these elements – renunciation, insight and resolution – can combine and somehow mingle in the sentence 'I will submit to my fate.' What a curious mixture!

'I will yield to evil': I will no longer resist the urge to do evil, I will not impose any restraints upon myself; I will give in to the terrible charm of evil and yield to it with a mixture of voluptuousness and horror – let destiny take its course!

'I will believe in God': This is very obscure. It is fairly clear what I mean by saying for example: 'I *wish* I could believe in God.' But what about *willing* to believe? Is this in my power at all? Perhaps I feel something rising from deep inside me and – at first obscurely and then with growing clarity – become conscious of a tendency at work inside me, slowly transforming my whole way of seeing things and leading me to believe in God. But then the question arises whether *I* or *something in me* wills: does the will come from me, as it were, or does it come to me from a deeper level of my being – my hereditary disposition for instance – perhaps even overpowering me? It is possible for me to defend myself against this tendency to believe and to look for reasons to invalidate it, as when I try to suppress a growing superstitious feeling. In that case there will be an argument and – sometimes – an end to the argument: either I overcome the temptation to believe or I say: 'Well, I have finally come to the decision to believe in God, I will no longer fight it.' In the latter case my willing contains an insight: by 'I have come to this decision' I mean: I have decided to acknowledge, to admit to myself, how deep the change is which has already taken place in me. It is like when I confess after a long inner struggle that I no longer love a being whom I believed I loved: making this clear to oneself takes *courage* – it is not a purely intellectual process – and the struggle to gain the insight, the overcoming of the obstacles in one's way, is what makes me choose the word 'will'. The same is true of belief in God. Instead of saying that I *will* believe in God, it would be better to say: I have decided to acknowledge that I already believe in him. This latter element – the collapse of my resistance and my decision to face that fact – is *my* will: it is not something foreign to me that comes up to me and gains power over me. It may therefore be my very own will to face my altered inner state, or to yield to the urge to believe which I feel inside me; but it may also be that there is something in me moving in that direction *against* my will, and that I observe in fear and with disapproval, even in horror, how this belief grows in me. How ambiguous language is! These very different processes are expressed in the same words: 'I will believe in God'.

'I will not live', like 'I have no will to live', does not mean 'I will kill myself', but perhaps: I am letting myself go, I no longer call upon the positive forces inside me, I no longer put up any resistance to the sweet temptation not to be. I abandon myself to the current pulling me into the abyss. It can also mean: I am tired, I am afraid of tomorrow, I have lost all curiosity and desire, I will shut my eyes and close my mind. Where is the willing in this case? I believe it is in my standing at a crossroads and making an inner decision: to reject all life-affirming, life-renewing forces, to put an end to all attempts to find a tolerable accommodation, to push away everything that could capture my interest and captivate me. And there is again that strange dichotomy: the will to stop living may come from *me*, as a rejection of the attractions of this world (like the ones Goethe represented at the beginning of *Faust II*), or it may come from deep inside me, and I may become aware of it only as a fully developed tendency in me.

'I will return to the past': I will not live in the present or future, I have turned my back on all that, I am looking back, I will not listen to the demands of the moment. Here again one shuts oneself off, steels oneself against all worldly attractions; and again there is the will to push them away, not to get caught up in them, to root out any remaining interest. But this will arises against the background of a sad and sombre mood without even a glimmer of hope. Without this mood the will would not be genuine. There are therefore experiences of willing *which presuppose a certain mood* as one of their essential features.

'I will not will': This sounds like a self-contradiction. For do I not *will* something, namely *not* to will? But this again means: I will no longer let myself be tempted to work towards something, to pursue a goal, to make an effort; I no longer believe in life; I will not take any interest in it. 'Oh, I am tired of life – what's the point of all that pain and pleasure?' But it can also mean something quite different, namely: I have gained the insight that you cannot rush things, that the important things in life cannot be won by force; so now I *will* no longer use my will; I will not force things, but wait for them or let them slowly grow and ripen in me. Such a refusal to engage in conscious willing and acting could also be couched in the words 'I will not will.' This will does not necessarily presuppose a sombre mood; for it can simply mean: I intend to stop willing for a while.

'I will' can also mean: I agree. 'I *will* the present international entanglements which are building up to a catastrophe' means: I approve of them, they correspond to my wishes, I do not protest against them, I find that this world deserves to perish.

4. AM I MASTER IN MY OWN HOUSE?

Are my thoughts at my command? Am I master in my own house? I can call up my thoughts; I can say: I will think about this; I can abandon myself to my thoughts and lose myself in them; but we often say that we let our thoughts 'roam': they carry us away and take us where they will – and, perhaps, where we ourselves do not will? Some thoughts torment us, and we try to shake them off like real nuisances. We close our eyes, move impatiently – we will not think of them. We do indeed succeed when we are awake and able to attend to other things. But when we are tired, ill, or about to fall asleep, they are back again. We try to think of something else, but we hardly take three steps in the new direction when we become strangely disoriented and find ourselves – God knows how – following the traces of our old thoughts. Or are our thoughts so wily that they use a harmless pretext, some association with what we *will* think of, to enter our consciousness? We drive them away, and the game begins again. There follows a struggle on my inner stage between myself and my thoughts: I try to get away from them and go to sleep, but in vain! They are back again for the nth time, ready to begin their cursed cycle. I am forced to watch them helplessly, even though I know that all this thinking is useless: I have gone through these arguments and counterarguments over and over again, and there is absolutely nothing new to be unearthed. And yet! My thoughts descend upon me; they are the pursuers and I am the prey; it is just like in the circus where the horses always trot along the same track; they will not let up before I have gone round the same track once again, weary and tired. And something like this happens to me in the waking state when I am under the influence of an emotion such as fear, anger, sorrow: I cannot suppress my thoughts: they spring as if from an invisible source and flood my consciousness. Am I, then, master in my own house?

Who has not been tormented by a popular song ringing in his ears like an annoying horsefly which refuses to be chased away? And what agonies we suffer once we start chasing after a forgotten name! We may not even be interested in the name because we may have turned to something else long ago – and yet an inner restlessness comes over us, and we keep on searching for it. We tell ourselves a hundred times that we care nothing for the forgotten name – but the very fact that it is forgotten turns it into a mischievous little creature which torments us from its hiding-place. We are off on an inner hunt: we stir up ideas, fragments of words or sounds, and discard them again. It is as if something made a fool of us, kept us in suspense indefinitely, as if to test us – a syllable, something about the rhythm, we know not what – but it will not slacken, so we give up looking for it, but are still haunted by the word we were looking for. Are we, then, masters in our own house?

And our memories! There is a kind of memory over which we have no power; it breaks a train of thought like a sudden flash in a river as it flows quietly on its way; I may be reflecting on the consequences of an assumption, and suddenly I see before me the image of a certain street, a square, a forest path, so clearly and in such detail that I could paint it; and in an instant it is all over, and the train of thought continues. What was it really? It is as if for a moment something had intervened, something alien and uninvited. Am I, then, master in my own house?

Let us now take another case. I am thinking soberly about something. But am I really the one who is doing the thinking? Far from it! My thoughts follow one another according to their own inner laws. Without my doing anything, one thought comes to be added to another related thought; I do not have the feeling that 'I am thinking them'. Could I actually intervene at will in my train of thought, could I slow it down at any point, or am I even free to think what I will? The simplest observation must show how illusory this is. I can – perhaps – divert my attention, and I can – perhaps – collect my thoughts; I can remove any physical obstacles standing in the way: I can retire to my room, lock the door, etc. But that is all. The first and most important thing I cannot do: I cannot make my thoughts arise at will. I put myself at their disposal; I am really only the stage on which they happen. According to Musil,

> the solution of an intellectual problem comes about in a way not very different from what happens when a dog carrying a stick in its mouth tries to get through a narrow door; it will go on turning its head left and right until the stick slips through. We do pretty much the same... And although of course a head with brains in it has far more skill and experience in these turnings and twistings than an empty one, yet even for it the slipping through comes as a surprise, is something that just suddenly happens; and one can quite distinctly perceive in oneself a faintly nonplussed feeling that one's thoughts have created themselves instead of waiting for their originator.[1]

And how is it when I make a discovery? It does not depend on me at all. I may have taken great pains to reflect on a problem and turned the difficulties over in my mind – but I cannot find a solution. Then – perhaps while I am sleeping restlessly or taking a break or getting up – I see it clearly without looking for it and know at once that this is *the* solution. Or the solution appears to me ready-made in the morning after a deep sleep, like Athena out of Zeus's head. Who can say how it happened? I do not even have the feeling that it was *my* doing; the thought somehow turned up unexpectedly – a visitor from another world: was it already somewhere before I had it? Or else I am thinking about a difficult question, get stuck, give up, and forget about it for years. And suddenly, without making an effort, I see the whole thing in a new light. I suddenly understand what always kept me back; 'the scales fall from my eyes': how could I have been so blind all this time and gone right past it? I marvel at how easy the solution is and at the same time at my own stupidity. Further, can one make a joke at one's discretion? Can one summon up an idea?

'The credit for that insight does not really belong to him. Various ideas just happened to come together in his head.' But why did they choose just that place? If no idea occurs to me today, I could put up a sign saying 'Not at home today'. But how can I tell whether I am not at home for the idea or it is not at home for me? I am tired and about to fall asleep when something occurs to me, a very interesting idea, and forces its way into my consciousness. 'Oh, at such an unusual hour?', I tell myself, 'Can't it wait until tomorrow?' No, the idea has very far-reaching consequences, threatens my entire system of beliefs from different sides, and expands rapidly; I *must* deal with it in spite of being tired and go into it now that I am fully awake. How impudent to disturb me in the middle of the night and condemn me to forced labour! As Kafka says: 'Now you come, bad thoughts, because I am weak; you want me to think you through right now. You are only interested in what is good for you. Shame on you! Come back some other time when I am strong. Don't take advantage of my weakness like this.' Am I, then, master in my own house?

How is it when I get a poetic idea? Sometimes I feel it coming; it is a painful pressure in my chest, as if I had been seized by something, and I strive to bring to the surface what I feel deep down. In other cases it comes quite unexpectedly. As I am shaving I look into the garden: a ray of sunshine falls upon a tree and weaves a gleam around it; the shading of the sky gives me a dreamlike feeling – and suddenly there are verses, like colourful shimmering bubbles from another world; oh, and sometimes it is only a garbled beginning, or a few words that trail off, followed by a stretch of fog and then again a clearing; and if I do not write the ideas down at once, they are gone – like trouts that got away, leaving me empty-handed and surprised at my empty hands.

And what is most remarkable, the rhythm of the poem trying to be born is completely fixed even before I find the words in which to clothe it. I listen to an inner voice, I am all ears, as if I were taking dictation – no, it is not quite like that: I see two or three verses, then it stops, but I have an inkling of how it goes on and try to conjure it up; sometimes I succeed and I throb with inspiration; sometimes it is all over: I try, choose, discard again, try to immerse myself in the inner vision – but the voice is silent. (How well I now understand what is genuine about a poem, what it owes to inspiration, how far the burst of inspiration carried it and what the poet added on to it. This clearly shows the seam where the two parts of the poem have been tacked together – pity the poor historians of literature who have no idea of what it is all about.)

Now are thoughts subject to my will? Am I master in my own house?

5. [CHARACTERISTIC MARKS OF WILLING]

We can distinguish very roughly between two kinds of willing: either the will is directed *directly* at what is willed, without the intervention or mediation of a third thing, as when I will to raise my arm for example. Or the will does *not* reach its object directly at all, but pulls a lever which, I believe or think I know, will bring about what is willed. The will then directly includes my picture of the causal connections between things; more precisely, the extent of what I call the 'will' depends on my ideas of the causal nexus in the world and tends to alter, shift, expand or shrink together with those ideas. 'I will teach this man a lesson', 'I will overthrow the government'. I can will this only if it is in my power; otherwise it is just an idle wish. To take an extreme example, if I believe – as many people did in the Middle Ages – that I can force God's will by using certain incantations and thus interfere with the wheel of fortune, my will can extend to practically anything; or more correctly, what I can will, or what I *think* my will can influence, will in that case depend on my idea of God's omnipotence, e.g. on whether I believe that he can undo what is done, etc. In that case I may even direct my will at the past and try to change it. Or I may believe that what happens in me, my experiences and thoughts, have some influence on what happens in the external world; not in the sense that, by being communicated to others, my thoughts may influence the course of events, but in the sense that their mere *existence* is somehow connected in a mysterious and inscrutable way with happenings in the wider world, that what is alive in me and subject to my will helps to change the happenings in the world. If I hold such a view, whether rightly or wrongly, the sphere of my will is very much wider than in the normal case. Think for example of the typical experience children have when they stand at a window looking at the drifting snow outside: they imagine that they can command the snow to fall at will; they tell themselves: 'Now a brief flurry, now just a few flakes; now quickly, now slowly.' This is what Freud calls the 'omnipotence of thought', which seems to be part of the world view of primitive peoples.

Let us take the opposite case: someone – say a patient who tends to imagine things – believes that he has a paralysed arm and cannot move it; this belief also cancels out his will; he can no longer even *will* to extend his arm; he does not think he can do it, and that is the end of his will. This shows how the will depends on our beliefs about what is in our power. However, things are not as simple as all that. For can he not at least make an *effort* to move the supposedly paralysed arm? This question is hard to answer. What does it mean 'to make an effort'? Is it to innervate the muscles? Or is it to set aside the idea that I cannot move my arm? Further, how can I tell that I transmit a stimulus from my brain to my arm muscles? Is there perhaps a specific experience of innervation which assures me of it? And how do I know that I really succeeded in my effort to set aside for the moment the idea that I could not move my arm, and how do I know that I am not mistaken about this? And finally: Does 'to make an effort' not mean as much as 'to summon up the will'? Or is there something more to willing, namely the actual movement? When I say 'The will is the act', is the act the actual movement of the arm, or is it, precisely speaking, the nerve stimulus travelling from the centre to the periphery? Here we have a further indeterminacy in the concept of the will.

In saying that he can surely make an effort to move his supposedly paralysed arm, I am presupposing that he can *try* to move it. 'To try' seems to mean less than 'to make an effort'; there is no reference to any energy in 'trying', unlike what is conveyed by the word 'effort'. Let us therefore put the question like this: *when* can he try? My heartbeat does not depend on my will; now can I try to speed up my heartbeat? Well, what am I supposed to do, how am I to go about it, in what direction should I concentrate my thoughts and my attempts? Should I perhaps try to listen to my heartbeat and – what else? I can perhaps *wish* that may heart would beat faster, and I may expect that if I concentrate on its beat it will suddenly start beating faster by itself – but is this what is meant by 'trying'? Can I try [*versuchen*] without trying anything in particular, or must I already *know*, at least roughly, *in what direction* I have to seek [*suchen*]? Can I try to move my ears if I do not know how it is done? I will perhaps pull all sorts of faces, expecting that if I pull it in a certain way my ear will move all 'by itself'. Is this what is meant by 'seeking'? Or can I speak of seeking only where I know the *way* which leads to what is sought? The concept of seeking has blurred edges, and hence also the concept of effort and that of willing. Can I will something only when I know 'in what direction' I have to will? Does the will presuppose a way? Or can I will blindly without the slightest idea of how to get to what is willed? However this may be, there can be no doubt that a patient who imagines that he cannot move his arm can nevertheless *try* to move it; for he knows very well – either from experience with his other arm or from memory – how to go about it. He as it were carries in his mind a pattern of the various bodily movements, and when asked to move his arm (or at least to try) he is in an essentially different position from the one we are in when asked to make our heart beat faster or to sneeze. There seems to be another element involved: it is also part of what is normally called willing that we receive afferent stimuli from the organ involved – and continuously so. My willing to grasp something with my hand presupposes that I feel my hand or 'have some feeling in it', that is, it presupposes a continuous flow of messages from my hand to my consciousness. If I lost all feeling in my hand, the question would again arise whether I could still speak of willing.

A person who has lost an arm can still have the experience of willing to grasp something with it; for the motor impulses as well as the afferent stimuli are still there; i.e., he still has a feeling in the arm which is no longer there. But if a person has been paralysed for years – can he still will to move his limbs? He can indeed give it a try and perhaps make an effort; but is that enough? Must he not also have *confidence* in his ability to execute the movement? This gives rise to borderline cases: some of the elements making up the will are present, while others are missing. The elements that are normally present seem to be the following:

(a) I have in my mind a pattern of bodily movement, so I know what to do when I will to move a limb. (What does this knowledge consist in?)
(b) I have some feeling in the organ, and continuously so, for stimuli flow from the organ into consciousness.
(c) There are motor impulses. (Do we have a specific experience of them?)
(d) I hold the belief that I can move the organ; 'belief' here can mean: I presuppose it tacitly, without having given it any further thought; i.e. it never occurred to me that I cannot do it; I am confident that I can do it; I sometimes tell myself, e.g. by way of encouragement: 'You can'.
(e) I experience a feeling of effort, of overcoming a resistance.
(f) I am aware of the movement I am executing as I am executing it, aware that my attention is focused on it. 'I know what I will'.
(g) I am aware that the will issues *from me*: I am not the instrument of another power which only uses my body as if it were a puppet to produce a movement.
(h) The movement is actually executed: I see it and have the corresponding muscular and kinaesthetic sensations.

In the light of this analysis I must modify the statement 'The will is the act': the condition it mentions is neither necessary nor sufficient; it is not necessary, for a person who has lost an arm can will to move that arm, even though the movement will not be executed (a, b, c, e, f, g);

and it is not sufficient, for I do many things automatically without our being able to speak of willing in those cases (a, b, c, d, h).

What is right about it is only that in many cases the actual execution is the criterion which distinguishes willing from wishing. But even this is not quite right. Suppose I say I will recall to mind the features of an absent person. If I do not succeed, it will be said that I only had the *wish*. But what if I succeed and really see the person before me in my mind? Am I to say that the memory image is the act? This would be very forced. And further, what if I say: 'I will now concentrate my thoughts on such and such an object'? Am I to call the thought process an act?

In listing the elements contained in the will, should I not have mentioned also that the will is *motivated*? This leads to the question: can I will something without a motive? Can I not intend to do something without having the least motive for it? I will to move my little finger for example and I do it; but I had no motive for it; I simply did it to prove to myself that I can do something without a motive. 'But you yourself have now mentioned the motive from which you did it.' Yes, that would seem to be a motive, though a strange one. But what if I move my finger because I just happen to feel like it? In that case there does not seem to be a motive; but is it a voluntary act? Supposing that I get carried away in a conversation and make certain intemperate remarks. Did I have a motive for it? Perhaps I did not know myself what I was going to say until shortly before I said it? Perhaps I did not myself anticipate the violence of my words? What really happened may have been the following: while speaking – and connected with what I was saying – I felt some agitation rising in me and swelling to the point where it discharged itself in words. That is all I am conscious of; now is this a motive? How is it when I yield to a temptation or, like impulsive people, to a sudden idea I am unable to resist – e.g. to pull someone's hair, to break into laughter at an inappropriate moment, to damage an object, or to do some other senseless thing? We then say: 'I do not know why I did it; it came over me suddenly, something pushed me to do it' or 'something drove me to say that.' Are these cases of willing? (The question of unconscious willing will be taken up later.)

6. WILLING AND WISHING

Is it part of the concept of willing that what we will is also what we wish? I can wish something without willing it, as when I deny myself a wish. But can I also will something without wishing it? Or does willing necessarily include wishing? It might be said: 'You will it; therefore you wish it.' Does the latter follow from the former? Suppose I am a judge and have to pass judgment on a man I happen to know and esteem personally; and suppose he has been found guilty and I have to sentence him according to law. In this case I do not wish to sentence him at all; I wish from the bottom of my heart that he were innocent and that I could waive the sentence. But I follow my duty and sentence him. Does that action flow from my will? Well, I hardly have a choice. Perhaps my will is simply not to go against my duty, and the sentencing follows from that decision 'whether I like it or not', that is, independently of my personal inclination. But can we still speak of an act of will in that case? Suppose I fail in my duty and release him – that would be an act of will: I shall have done something, overcome a resistance, i.e. the pressure of my conscience, and expended some effort – in short, I shall have willed. But now suppose I go through a certain struggle, am pulled back and forth between opposing forces, and finally make up my mind to turn a deaf ear to the voices of temptation advising me to release my friend and instead go ahead and sentence him – I shall again have willed. But if I simply do my duty without any inner opposition and act only as a judge should – there will have been no willing. If I decide to sentence him, my willing goes together with my wishing; if after some inner conflict I decide to sentence him, the willing runs *counter* to the wishing. Or does it? Could I not say in that case that while I had the wish to see my friend released, I had another stronger wish, namely to be faithful to my sworn duty, and my decision followed that stronger wish? In that case my willing would have been in accord with my wishing.

Now how am I to decide *what* I wished in that case, whether I had one wish or two wishes. I *may* go through such considerations as: 'I wish to preserve my self-respect; if I were to release him, I would lose it or invite a disciplinary inquiry, etc.' But in general it seems to me that we do not have such thoughts. The decision I am faced with is not whether I would prefer to release my friend or keep my conscience clear. Rather, the fact of the matter is that there is on the one hand the inexorable 'thou shalt', my duty, and on the other hand my inclination. I do not seek to weigh one wish against another, but I feel a conflict between wish and duty. That I would like to keep my self-respect, or that I am in danger of losing it, is an additional idea which is not immediately in the forefront of my consciousness. I do not ask 'Do I wish to do something which I know will cost me my self-respect or have other unpleasant consequences for me?' The question is: '*May* I do it?' This *primary* thought may be accompanied and reinforced by *other* thoughts and wishes – but they are not essential. My dominant thought is whether I should do what I would *like* to do or what I *ought* to do. Considerations concerning my own person are not very relevant here; they *may* come into play, but it seems to me to falsify the facts to say that we are here dealing with a conflict between two inclinations – an altruistic inclination towards my friend and an egoistic inclination to preserve my own integrity. If this were correct, I should deserve to be morally condemned – namely as an egotist – if I fulfilled my duty as a judge, and to be morally commended if I neglected it. This consequence sheds light on the interpretation that in fulfilling my duty I am at bottom only following my inclination.

Willing can therefore go against wishing. I can will something I wish, and I can will something I do not wish.

Can I will to wish something? Can I will wishing itself? I think it is meaningless to say: 'I will wish', except perhaps in such phrases as: 'He is at the door; let him in: he will wish you all the best', which means something like: 'He will offer you his congratulations'. In everyday speech we can indeed say at a pinch: 'I will wish for such and such a thing' as a paraphrase of 'I wish for such and such a thing', perhaps with the incidental thought that the wish is not yet quite definite. But apart from such improper uses it makes no sense to direct the will to wishing. The will can have different things as its object, but never a wish.

Can a wish be directed at the will? Can we say: 'I wish to will'? Suppose I suffer from weakness of will in the sense that I find it difficult to will anything; I could say about myself: 'I wish I willed' in the sense of: 'I wish for some will power'. Thus a wish *can* have the will as its object.

Can a wish have another wish as its object? Can I say: 'I wish to wish' – or is that nonsense? We do not usually express ourselves like that. And yet I can imagine myself saying: 'I do not have any wishes; that is precisely what is so sad about my condition; I wish I could wish or have some wishes'. Is that a self-contradiction? I am saying in one and the same breath that I do not have any wishes and that I nevertheless wish for something, namely to have wishes. But let us distinguish more clearly between a concrete wish directed at the fulfilment of some desire, e.g., 'I wish to take a nice trip' – let us call this a first-order wish – and a wish directed at the occurrence of wishes – a second-order wish. Now if I say: 'I do not have any wishes; I wish I had wishes', what I am saying is, strictly speaking: While I do not have any first-order wishes, I do have a second-order wish – and now everything is all right.

Can willing take aim at itself? Can we say: 'I will will'? No; in this case willing seems to take a run-up and – land in the void. When I will, I will *something*; but this something cannot be willing itself.

The logical grammar of the words 'wish' and 'will' thus seems to point to the following rules.

> From 'I wish A' it does not follow that 'I will A'.
> From 'I will A' it does not follow that 'I wish A'.

It makes sense to say:

> I wish A and I will A:
> I wish A but I do not will A.
> I will A and I wish A:
> I will A but I do not wish A.
> I do not wish A and I do not will A.

It is permissible to say:

> 'I wish to will',

but not:

> 'I will wish'.

It is permissible to say:

> 'I wish to wish',

if one observes the difference in type, but not:

> 'I will will.'

7. WILLING AND KNOWING

We have said that there is willing only where there is some resistance being overcome. The effort connected with it is part of what we experience in willing. But this presupposes that we *recognize* the resistance and become conscious of it. To stay with the simple example of grasping the knife held firmly in place by the magnet, if I did not know that there was some resistance, I could not will to grasp the knife. And could I will to grasp it if I had noticed that it could not be picked up, but had no idea how to overcome the resistance? Suppose I found myself in a fairyland where the most unexpected happened: when I try to pick up the knife by force it recedes into the distance or vanishes altogether, but when I happen to touch some other object I can pick it up without any effort; if I do not have the least idea of the laws governing this fairyland, can I really will to grasp the knife? I can indeed wish to grasp it; but willing includes having an idea of how to translate the will into action. *So willing presupposes some insight.* It does not presuppose that I know *exactly* how to reach the goal, but only that I have a rough idea of the direction in which I must go. We can say: 'I willed to do such and such a thing; I first tried it this way and then that way; but I did not succeed'; and that is not a contradiction.

So willing calls for some insight, but not for too much insight! When we will, we are in a peculiar situation: we see what we will; we also see the means by which we can attain it, though perhaps a little less distinctly; we also foresee some of the immediate consequences, if only with probability; and everything else gets lost as in a fog. This remarkable situation, in which we have a clear view only of the next few steps ahead of us while all the rest is hazy, is extremely significant for willing. For suppose it were otherwise: suppose we could survey the consequences of an action into the most distant future with perfect clarity, the ones we approve of as well as the ones we disapprove of, we should stand there unable to move, as it were; we could not make up our minds because in most cases the pros and cons would cancel one another out, or because we should have a hard time making a decision being conscious of the fact that by acting on it we would be sacrificing such and such a thing, which mattered a lot to us. Acting and willing are possible only because the future is hidden from us: the will requires uncertainty. If I were completely ignorant, I could not will; and if I were omniscient, able to look into the most distant future, I could not will either. *Willing is possible only in the grey area between knowledge and ignorance.*

The same thing is shown, incidentally, by an experience we are all familiar with: going too deeply into the possible consequences of an action tends to weaken the will. This leads to the question: What is weakness of will? And what is strength of will?

First, what is intensity of willing? Is it something we experience immediately? Is it the violent, impetuous, passionate aspect of the will? Or is it perseverance, the existence of which can be demonstrated only in retrospect? If we disregard perseverance, there are two different elements which make the will appear strong: first, the willing subject sees the action and its consequences more or less distinctly before his mind; he sees all the difficulties, but does not let himself be frightened off and overcomes the obstacles with, as we say, great will power. The resistance being overcome is *a measure of the strength of will*. Secondly, the subject's field of vision is so narrow that he sees only the one thing to which the will is directed while all other considerations recede into the background. This happens in a passion and is one of its most characteristic traits. A person in a passion is as if intoxicated: he loses perspective and all sense of proportion; a huge wave sweeps over him: all other things, including any thoughts of consequences, are as if washed away for the moment, and his entire being is filled only with the one all-absorbing thing.

Oddly enough, a person who wills intensely in this way is actually more passive. We say of such a person that he *abandons himself to his passion*. We could say either that he *wills very intensely* or that he *has no will of his own* with respect to his passion and that his passion wills *in him*. We could thus distinguish between the person's willing and his passion's willing; the passion robs him of his will and at the same time fills him with an intense will. A person is robbed of his will in so far as he is no longer able to see things clearly and soberly in their mutual relations and to shake off the powers that control him. His willing is intensified at the cost of his field of vision, which becomes narrower. Further, his will is intensified only with respect to his one oversized, overemphasized goal; in any other respect his will is not intensified; whereas a strong-willed person acts in *any* situation with a clear vision of his goal and a knowledge of how to overcome the difficulties. Intensity of willing can therefore mean two different things: a strong will and a passionately aroused will.

Now each of these two states has its own opposite. The opposite of a strong will is a listless, tired will which can no longer muster up the energy it takes, even though the subject may see that there is an urgent need for action; in this case there is reluctance to face the difficulties, a feeling of not being up to the struggle, submissiveness, lethargy. The opposite of a passionate will is a kind of indecisiveness in which the subject never gets down to action because of doubts and worries; he takes pains to think out all eventualities, considers all the pros and cons in never-ending detail, and as a result is unable to decide on anything.

We could say that in this case the will slackens *because* the field of vision widens.

The two kinds of weakness of will may, incidentally, be combined: someone who feels lethargic and reluctant to act may seize on the *pretext* of looking at the action and its possible consequences in an extremely critical light and use imaginary difficulties as an excuse for inaction.

The opposite of *perseverance* is *inconstancy*. Considerable strength of will may be accompanied by inconstancy: the result is one of those impulsive characters who get carried away easily but are unable to stick to a definite goal for any length of time, who are always open to new impressions and let their actions be determined by them. There is something else in this impulsive willing: the feeble contribution of reason in determining the will, the subordination of the intellect and of the critical spirit; which is why impulsive people rarely think critically and the critically minded rarely act impulsively. The result of strong impulsive willing is similar to that of weakness of will, in that different impulses often cross one another and cancel out in the overall result: such people will find that they 'did not get anywhere' in life.

Intensity of willing is, incidentally, subject to fluctuations: a weak-willed person may on occasion rouse himself to perform amazing acts of will, and an energetic person may give the impression of complete weakness of will, e.g. in case of a conflict of motives. Physical condition, fatigue, worries and fears also come into play. Fear or anxiety can, incidentally, play a double role: it can spur a person on to perform a tremendous act of will he would otherwise be incapable of; or it can leave him paralysed like a rabbit hypnotized by the sight of a snake and unable to budge. What determines whether anxiety stimulates or paralyses? A related fact is that some people become cool and calm in the face of danger and think things over in a flash, whereas others 'lose their heads' and rush headlong into it. It thus seems that anxiety can either paralyse or spur on the will as well as the intellect. The same would seem to apply to worry and pain. Cf. the expressions 'petrified with pain' and 'going wild with pain'.

The cases of the will paralysed by passion and indecision or anxiety show that action involves some insight as well as some ignorance and that too much clarity destroys the possibility of action.

But is it only uncertainty which makes action possible? Is it not rather a certain *belief* or *confidence* that the action will have good, desirable results? Surely the latter! But in that case the will thrives in an atmosphere of positive belief rather than intellectual ignorance and inability to predict the future. My acting or willing must be accompanied by a certain belief: that what I will is 'the right thing' and will 'turn out well'. If I entertain too many doubts and scruples, I destroy this belief and weaken the will.

Here we get a glimpse of the connection between a theoretical world view and a practical attitude to life: scepticism goes hand in hand with a certain lack of will power, whereas a positive world view, no matter what its content, tends to intensify the will. Or rather, those endowed with a fresh and lively will tend to be drawn instinctively to a positive world view. There is probably no point in worrying here about cause and effect – from what has been said we can, it seems, conclude that there is a *correspondence* between a kind of willing and a kind of thinking.

I said that willing involves a certain belief. Now belief has two opposites: doubt and knowledge. Doubt creates a situation in which willing is inhibited. Now what is the influence of knowledge? If I were all-knowing, so that I could foresee all the consequences of my action, could I still will in the ordinary sense of the word? Or does knowledge also destroy willing when it supersedes the instinctive belief connected with willing? I do not now want to speak of the *indecision* which comes with such knowledge and is simply due to the fact that any action will have undesirable as well as desirable consequences, but I want to consider the *effect of replacing the belief* by knowledge. Thus when I act or will something, my willing is carried by a kind of confidence that it will turn out well. Willing takes place, as it were, in an atmosphere of faith. (This is not always the case, though: if I make up my mind to die, it may be because I have lost the last shred of faith.) But if I no longer have this belief because I know exactly what will happen, how is this going to influence my willing? I am no longer shrouded in that atmosphere; I have no expectations, no hopes or fears, no curiosity; I see clearly and distinctly what will happen, and I see myself doing such and such a thing – but what kind of state did I just describe? If I know all that in advance, it will be impossible for me to will anything. For willing presupposes that by doing something I change something in the world, that in deciding to do something I decide between different possibilities, and that this decision lies exclusively *with me* or comes *from me*. After all, willing is a kind of choosing, and choosing makes sense only if the choice I make changes something just because I make it. Of course if I know everything in advance, including my own decisions, I can no longer believe that I can change or decide anything – for everything has, as it were, been decided, and my only role is that of a purely passive spectator who watches events glide past as on a movie screen and to whom even his own actions must appear like cinematic happenings, almost like a dream. In other words, if I know everything in advance, I have no longer any *motive* for acting, and without a motive there is no longer any willing. It thus appears that willing requires *intellectual uncertainty* after all and is impossible without it.

What I have said sheds light on the question whether I can foresee my own behaviour. If I could foresee it exactly, I would not have the feeling that I was a willing subject. If there were laws in psychology making it possible for me to calculate in advance what I shall or shall not do, I should be in an extremely odd position: this knowledge would destroy my ability to will. For the knowledge that I shall do it stifles the decision to do it. In reality I always have the feeling that there are different possibilities and that I can decide between them: I can do either this or that or the other thing. I do not know what I shall do before I have done it, and then only on the basis of what I have done. I am in the same position with respect to other people: I can perhaps exclude some possibilities, regard others as improbable and hesitate between still others. That is, the knowledge we have enables us to estimate – roughly – the limits within which a person's behaviour will occur. The possibilities form, as it were, a cone of dispersion which is open towards the future. That is, we can have a general picture of the possibilities open to a person without being able to say what exactly he will do. In this sense we could say that whatever laws there are in psychology are *limiting* but not *determining* laws. I, too, always have ahead of me a cone of dispersion of my future possibilities. As I travel into the future, I take this cone along with me. That is, the future always appears to me in the form of such a cone – as something that is still indeterminate in some respect and will be determined only through my action.

Thus knowledge of the future has two different effects: on the one hand it makes me hesitant, for the farther I can see the consequences of my action, the more difficult it will be for me to reach a decision, and the more persuasive I shall find the reasons for and against each individual action; this will finally lead to a kind of paralysis of the will. I can no longer will because the reasons for and against cancel one another out. But if knowledge of the future is conceived in such a way that it includes also my own decisions, the concept of willing ceases to apply, for it no longer makes any *sense* to speak of willing. In the former case, considering everything I know, I find it *psychologically* impossible for me to bring myself to make a decision; in the latter case it is *logically* impossible to will: the idea of willing is no longer applicable.

Thus willing requires a certain haziness: it requires that we see a few steps ahead, but not too far, and that we do not know exactly how we shall decide in a given situation. The latter remark invites several objections. Are there not many situations in which I know with certainty how I shall decide? If someone asks me to join him in committing a crime and I am a decent person who is not at the moment a prey to his passions, I know with certainty that I shall decline. Does this make my willing illusory? Was I not therefore too demanding in saying that I must not foresee my own decisions? But if I decline flatly to be an accomplice to a crime, the question arises whether this was really an act of will: I did not wrestle with myself before deciding not to do it; I did not first feel and then suppress a temptation; in short, what is missing is precisely what is characteristic of willing. From an external or legal point of view my refusal may look like an expression of my will; but seen from inside I did not will; nor would I say: '*I have decided* not to participate in this crime'; I would say this only if I had contemplated the idea, entertained the possibility, and then dropped it. In other cases I *think* I know in what direction I am going to decide, but I do not know it for certain, as shown by my being occasionally surprised at the way a decision turns out. I then say: 'I should never have thought I would do that' or 'To my surprise I suddenly decided to do that'. In this sense we get to know ourselves only afterwards through the decisions we have taken. As Schopenhauer says, 'through that which we do we only find out what we are.'[2] Thus in making a decision, I must be in the following state: I hesitate, if only for a short time, between different possibilities; my attention is divided between conflicting ideas; I have the feeling that it is up to me whether I do this or that; in short, there must be a *conflict* whose outcome I do not foresee with certainty. The more distinctly I foresee what I shall do in a given situation, the less room there is for willing. So to that extent willing is tied up with *inability to foresee* one's own decision. In willing I must have the feeling that *I* decide and that I could also have decided otherwise, for otherwise it would not have been willing.

There is, incidentally, another confusion that comes into play: the failure to distinguish between will and desire. I may possibly be seized by a strong desire – like a drunkard at the sight of a bottle – and I may even know this in advance with certainty; this foreknowledge does not affect the strength of my desire in the least; but this desire is not the product of my will or the result of a decision, even though the desire may make me will things I should not have willed otherwise.

I could also will if I had known the result of my decision in advance, but had forgotten it at the time of willing. I must have the feeling that I can do what I will *at this very moment*.

8. ON THE INDETERMINACY OF WILLING

We are all familiar with the experience of not being sure whether we willed or not. We *do* something; but the question whether we willed what we did leaves us groping for an answer. Suppose I am in a bookstore; my eye falls on a book, I feel attracted to it and I buy it. Did I will it? Good heavens! what can I say? I cannot deny having bought the book 'of my own free will': nothing compelled me to buy it; and to that extent it was 'my will'. But if I were asked 'Did you *decide* to buy it?', the only truthful answer would be 'No'; for the decision presupposes that I already had the goal in mind beforehand, that I had the plan or intention to purchase this very book, whereas in reality I yielded to a momentary impulse. Even if I were asked: 'Were you determined to buy it?', my answer would have to be in the negative. Being determined means that my will is so strong that it would overcome such and such obstacles, or that I would be ready to make such and such sacrifices for it, etc. Was I determined in this sense? No; and yet I willed; we can thus will something and do it without being determined to do it.

We could distinguish between 'willing something' and 'having decided to do something'. The latter says *more* than the former: it includes willing, but adds that I had *thought beforehand* of what I had willed and, in addition, that I had already made up my mind to do it at the first opportunity. For I could have contemplated what I willed, merely toying with the idea of doing it, or just thinking about it before making up my mind to do it. There is thus a wide area between planning and deciding which could be described by using the vague expression 'having the intention'. 'To have the intention to take a trip' may mean: to direct one's thoughts to it, let them roam over it, focus on it, make plans, think about means to that end, consider consequences – and one more thing! For what I have described so far would not yet be *having an intention*, but merely considering an idea or entertaining a possibility. What must be added to make it an intention is the *wish* to actually take the trip. As long as I see it only as a theoretical possibility and feel no desire to travel, nothing to urge me on – what is supposed to distinguish having the intention from merely thinking of it or even being afraid of it? The intention thus includes the wish; in fact, we use language in such a way that in many cases the words 'I have the intention' mean virtually the same as 'I have the wish' or 'I wish'. For example, 'I wish to rent a room' is equivalent to 'I have the intention to rent a room'. True, the meanings of these two expressions do not coincide completely; the 'intention' to rent the room suggests, as it were, that I really have that intention, that I am serious about it; whereas the 'wish' points to a much wider and vaguer area, as when we speak of wishes in the sense of roaming desires, playthings of the imagination we are not serious about.

The intention does not *always* include the wish: I can say that I have the *intention* to do something even when I have no such wish, and even when it goes *against* my wishes, because my action is dictated by the realities of life rather than by my wishes.

Thus when I say 'I have decided to take a trip', this presupposes that I had the intention to travel and had thought about the trip beforehand; but I am saying still more than that: something that accounts for the difference between intention and decision. How am I to formulate that difference? It will be noted that the word 'decision' suggests willing, and we have seen that we speak of willing only where a certain resistance is overcome. Thus we characteristically say that I finally came to the decision after 'wrestling' or 'struggling with the problem', or that I 'forced myself' to make the decision – turns of phrase which illustrate a conflict between several imagined goals, or between an imagined goal and a state of lethargy. It will not occur to anyone to express his intention to go out by saying 'I have decided to go out' unless it is dangerous to go out because there is some shooting going on outside or because it is forbidden, etc. Once we overcome an internal *resistance*, we can speak again of decision. And once we have made the decision, we can again be asked how firm it is, whether we can easily be made to change our minds or not, etc. Language takes account of this by placing different expressions at our disposal, such as: I am willing, I am more than willing, I will, I definitely will; he is determined, he is fully determined; he is firm, tenacious, unyielding, unshakeable in his determination; he is unwavering in his resolution; etc.

We could accordingly form a series whose elements hang together like the links of a chain and go from the weakest to the strongest: from toying with an idea far from reality to being absolutely determined to realize it.

I consider a possibility; I toy with the idea; I feel a desire; I have the desire; I entertain a wish, I have the wish; I have the intention; I am ready and willing; I am determined; I am fully determined.

Let us see how toying with ideas turns into a wish, a wish into intention, and intention into decision. The point in time at which an intention – which is still almost a wish – changes into willing is called the 'moment of decision'. While making a decision to do something is a process we can observe in ourselves and fix in time with more or less precision, 'having decided' to do it is a state in which we find ourselves but whose presence in us we cannot ascertain continuously; rather, 'I have decided to do it' is similar to 'I know the alphabet' or 'I can play chess', and like these formulae it expresses a readiness, a disposition. 'Having decided' is not a state like 'having a temperature of 100° Celsius', of whose existence we can convince ourselves continuously by means of a temperature curve. There is no 'decidedness curve', not because we lack a suitable apparatus for measuring it, but because having decided is *grammatically* a state in another sense of the word. This is shown, among other things, by the fact that it makes no sense to ask: 'When you have decided to study, are you decided from morning to evening without even a second's interruption?' We can speak of a person 'wavering' in his decision: he makes it, gives it up again, changes it, and makes it again; but having decided is not a mental state whose existence can be verified continuously by introspection. Let us go back. I began by remarking that there are cases where we do not quite know whether or not we willed something. What is certain is that we can will to do something without having decided to do it; and these are cases where we act, as we say, 'on the spur of the moment', without thinking beforehand of what we do. But then there are other cases where it becomes really doubtful whether we willed. Suppose I let myself be drawn into doing something I do not really quite approve of, e.g., playing a prank on someone. What should I say if I were asked afterwards 'Did you will it?'. I did not have a prior intention; I would not have thought of it on my own; but I *did* it; actually I joined in only in order not to be a spoil-sport; did I will it? Strictly speaking I should have to say: I did not make enough of an effort *not* to do it; I did not will the contrary hard enough. I am, of course, *liable* for what I did, legally as well morally: the consequences must be put down to *my* action and *my* will; but psychologically it would be difficult to interpret my lack of a contrary will as a will. Now suppose I give in to a sudden impulse and do something: did I will it? The answer is again: I willed it in the sense that I did not try hard enough to suppress the impulse, and I am responsible for my action. (It will be seen that in this case I am being drawn into it from within.) But now, such cases are very frequent, and it is common knowledge that we do not credit impulsive people with much will power, but are rather inclined to say that they find it hard to resist their impulses. ('A passing fancy.')

There is a similarity between impulsiveness and passion: when we are very excited, angry or afraid, or for that matter enthusiastic, our normal limit of resistance is lowered, so that we do things or, better, do not resist things we should resist in a normal frame of mind. Under the influence of a strong emotion we seem to be able to perform great acts of will because the contrary will is weakened. In such a state the area of real acts of will decreases, whereas from the outside it looks as if the will was strong. Thus in a case of passion it is also difficult to say: to what extent did I really will it? There is also a similarity with being intoxicated.

You will notice that in judging such cases there is another thing to be taken into account: namely the *attitude* we adopt afterwards to what we did. Maybe I got carried away momentarily without thinking – whether by the company I kept or by an impulse I could not resist or by passion. But there is also the question of how I judge my own action after regaining my composure: do I recognize it as something springing from *my nature*, something I agree with and am certain I would do again under similar circumstances, or would I like to take it back – by saying for example: 'I cannot understand what got into me; I cannot have been my normal self; I must have been beside myself'? In a word, it all depends on whether I *acknowledge* what I did. If so, I did will; and if not, I really only failed to will the contrary, even though I am, of course, liable for what I did.

Now what if I do something, say in an emotional frame of mind, and then ask myself: did I will it? In one sense I did not will, for I did not have the prior intention to do it; but I did do it, and now that I think about it something in me agrees and something disagrees with what I did. Did I will it? *I am at odds with myself.* And that is all that can be said about it; what it shows is how the concept of an act of will gets blurred also in another direction. As Nietzsche says: 'The criminal is often enough not up to his deed: he extenuates and maligns it.'[3]

9. THE DIVIDED WILL

The preceding discussion has called attention to the phenomenon of *being split in two*. We are often split in two in willing: we will with one part of our being but not with another. When we do something in such a situation, we have to overcome an *inner resistance* (Macbeth). This leads to the question: am I always split in two in willing? Or is there something like unbroken willing, where we will with our whole being? In other words, is willing always accompanied by counterwilling; is it always the result of two opposing forces? It seems to me that there is a kind of tough, dogged willing: we pull hard like a horse in harness, with clenched teeth, steeled against and deaf to anything that would make us pull back from it. We are like a captain on the bridge when his ship is in distress: fixed, immobile, focusing on the one thing, and oblivious to everything else. We speak of willing because here too there is some resistance, though *only on the outside*. This yields a particularly pure and unadulterated form of willing.

On the other hand, if there is some inner resistance, if we will only with part of our being, the will is somehow broken and divided and becomes, in fact, weaker and less certain, for the countercurrent breaks through at the opportune moment and cuts across the willing: this is what gives rise to uncertain and erratic willing, to not being forceful enough and hence failing to act, or to taking a false step at the decisive moment; what we know as failure often has its roots in that we do not get wholeheartedly and unreservedly involved. Objections, doubts and counterarguments render our gaze uncertain, so that it is no longer fixed steadfastly on the goal, but wanders off to one side or the other.

Whether a person wills as an unbroken whole or uses only part of his will power depends on the situation and the person. We all know situations where we do everything in our power to reach a goal or to avoid some danger; where we proceed single-mindedly. And there are other circumstances in which we have to force ourselves to make a decision and then pursue the goal only listlessly. How often we find ourselves in the latter situation depends on us, on our character and temperament, but also on our field of vision. For action is possible only in a certain half-light; strictly speaking, everything we set our sights on is so complex and multifaceted, and fraught with so many consequences, some of them desirable, some undesirable, that if we saw all these consequences spread out before us, and if our eyes were sharp enough to penetrate the obscurity of the future, we should, in the end, be in the position of the millipede: we should no longer know what to do. The more clearly, the less dispassionately and the more objectively we conceive a situation and follow up its supposed consequences, the more doubts and difficulties will arise before us and the weaker will be our drive to act; whereas a certain narrowness of vision, say staring as if spellbound at the one goal before us and not seeing anything else, will strengthen the will; which shows that a theoretically minded person who is used to seeing all sides of a question and to weighing the pros and cons is not very good at practical action, whereas a man of action will hardly have any clear and unbiased views about the nature of things. Like Napoleon, he may of course be an excellent judge of people and things. But he must be able to narrow his vision at least temporarily. As Nietzsche says: A sign of strong character, when once the resolution has been taken, to shut the ear even to the best counter-arguments. Occasionally, therefore, a will to stupidity.'[4]

In a passion the field of vision narrows; any strong but dubious conviction has the same effect, as when I am absolutely convinced that I am right. Scepticism widens the view and therefore leads to indecision.

10. WILL AND EGO

Schopenhauer says: I can do what I will; but can I also will what I will? Will what I will? Of course! Whatever I will is what I will; that is a tautology. But there is another sense, which is probably what Schopenhauer had in mind. In that sense a person's will is his real self. For he himself is the way he wills, and he wills the way he is. Hence to ask whether he could will otherwise than he wills is to ask whether he could be someone other than he is; and that he does not know.

I can do what I will: I can, if I will, give everything I have to the poor and thus become poor myself – if I will! But I cannot will this, because the opposing motives have much too much power over me to be able to. On the other hand, if I had a different character, even to the extent that I were a saint, then I would be able to will it. But then I could not keep from willing it, and hence I would have to do so.[5]

The question 'Can I will what I will?' would then mean: can I be a person who can will such and such a thing (which I cannot now will)? And can I will to be another kind of person? Can I choose my character? This leads to the question to what extent my character is imposed on me from without (by heredity and upbringing) and to what extent it is my own doing. To settle this question, we should have to go into the concepts of the ego and one's personality and observe how the external world impinges on the ego.

What does my feeling of doing something 'spontaneously' consist in? How do I know that what I do flows from my will and that I am not mistaken about it? Well, perhaps I remember what made me do it, I recollect the motives from which I did it, etc. Is that all – and does it rule out any possible mistake? Suppose I am acting on a post-hypnotic suggestion, thus following someone else's will, but without knowing that I am under such an influence. Is it possible for me to discover that what I am doing is not my own doing, even though physically speaking I am the one who is doing it? If someone asked me why I am doing it, or if I myself raise this question, I shall not be at a loss for an answer, but mention without a moment's hesitation a motive I now *believe* led me to do it. Suppose that under hypnosis I was ordered to open my umbrella in the room at a certain time. Asked by an astonished onlooker why I did it, I may perhaps reply: 'Oh, it just occurred to me that there is a hole in the umbrella, and I wanted to see where it is'. This sounds quite plausible. Do I have a way of finding out that I did not open the umbrella on my own initiative? I believe this question cannot be settled by mere introspection. No matter how carefully I look into myself, I will not find anything to distinguish this process from other actions 'willed by me'. Only when I come out of my amnesia and am able to remember what happened during the hypnosis, or when I am told the whole story afterwards, will it occur to me that I was not really 'the doer of my deed'. There is no inner criterion for distinguishing a willed action from an action performed under someone else's will if that foreign will is beyond my ken. So the statement 'I willed it' based on my self-consciousness is *not absolutely reliable*. What is decisive is, rather, the observation that my action does not quite match what I usually do, that it does not fit in very well with the rest of my behaviour, that it somehow 'sticks out', that some of its features are not sufficiently motivated and do not resist critical examination – unless my statement is taken simply as a report of what happened during the hypnosis. In short, what is decisive is certain reasons which go far beyond what I immediately experienced at the time I performed the action and what I remembered shortly afterwards.

This seems to bring out a remarkable fact: that I cannot be absolutely certain that what I am now doing is really willed by me; that the data now at my disposal are not sufficient to exclude every possibility of my being the instrument of a foreign will; and hence, that the statement 'I willed it' is not an ultimate incontrovertible truth.

So far we have considered the possibility of my being under the influence of a foreign will. But the matter becomes even more entangled when we consider that there is really no agreement on what constitutes 'my will'. These are not ill-conceived questions arising from philosophical scepticism. They sometimes arise in daily life, as when we are moved by a strange impulse. We then ask: 'Where on earth does this strange impulse come from – it does not fit at all into my conscious waking personality? Does it really come from me? Or does it come from my subconscious? Or is it something that rises out of the depths of the past and comes over me?' In asking this question, we act as if it clearly made sense to say that an impulse 'comes from me' or 'my own nature'. But what does this really mean? Think of the things that enter into my will! Can I clearly demarcate my will – can I draw a sharp boundary between myself and whatever survives of my ancestors in me? And when I will something and make a decision, how do I know that other beings who preceded me did not join me in pulling the lever? Perhaps I come closer to the truth if I say that I am a collective? Perhaps the idea of an isolated individual is just a fiction?

But is there not something rational in willing, something I approve of, which arises from reflection and makes me say that it is I who wills? Perhaps there is such a thing; but there is no absolute guarantee: for an inner impulse or urge can *masquerade* as a rational will, by putting forward quite plausible reasons. I can approve of these reasons; I can *believe* that it is I who wills – and yet the impetus to all this comes from something totally unknown to me. This is a typical case of 'rationalization'.

The situation is thus much more difficult: not only am I unable to decide with absolute certainty whether my action springs from my own will, but the question does not even seem to have a clear meaning. It could be said that one's ego gets blurred inwardly in the direction of one's ancestors and is perhaps also indistinct in some other direction, and this is why the question raised does not have a clear meaning. Let us imagine a clan which outwardly appears as a unit, only much more tightly knit than any human family. Imagine that the members of this clan are as closely attached to one another as the individuals of a coral colony, with the same feelings and emotions running through them all; and imagine that this clan expresses a will in which the volitions of individual members are blended together indistinguishably. If we now replace the spatial picture by a temporal one, that of a succession of people extending over different generations and yet extremely closely attached to one another – would this not be a good approximation to reality, a good picture of the structure of a human being? If we adopt this view, we are no longer able to say clearly what belongs 'to my will'.

This makes it clear, not only that we sometimes do not know where a volition 'comes from', but also how we can get the impression that we are not responsible for what we do, but merely execute the orders of a foreign will. Suppose I have the following experience: my limbs move as if pulled by an invisible force; I have a strange dreamlike presentiment of what I am going to do the next moment, and in doing it I also have a certain feeling of effort. Now would this be identical with the experience that this action flows from my will? That is, would everything be exactly as if I had done it and as if what I had done had issued from my will? Or could I still hesitate and say: 'Strange, it does look like it was *I* who willed, but I feel it was not me at all. It is as if my body was used only as a means by which something else willed'? Perhaps I have only different partial experiences of willing without really willing? There is something eerie about this experience: we feel as if we are being forced out of our own ego, and we have the sense that someone else is willing and acting on our behalf.

I see that my will is not something tangible, sharply definable: it dissolves before the mind's eye into a cloud of possibilities – of partial strivings, feelings, dispositions and incipient movements, some more central and others more peripheral, all of them lying in wait for it, so to speak. To use a simile: I am like a cloud, without sharp outlines, massing, swaying, stretching this way and that, changing shape, sometimes evaporating, sometimes forming, playing an eternal game. *And this is how my will is.* That is, the will in me merely mirrors what is shapeless, vague and fluid in my nature. In the act of will this indeterminacy takes shape for a moment and solidifies – some part of me gets through to reality: a decision is a step towards self-realization. The peculiar indeterminacy I feel before the action – the feeling that I could do the one thing just as well as the other – does not rest on an illusion: I really feel the different possibilities of which I am made up, one of which is about to be realized. I cannot foresee my will because this would require data which will be available to me only in the future when the decision will already have been made: the moment *before* the decision I am still immature, as it were; something in my nature is still open and will *mature only in the decision*.

Thus in the different acts of will we perform in the course of life, certain possibilities that lie dormant in us step out into objective reality, as it were, and we ourselves gradually assume a *definite shape*. To will and act is to go from being shapeless to being shaped. When we perform an action, we realize not only it but also ourselves, or part of ourselves; whence the deep feeling of the importance of an action: it is not only that we change the external world by what we do, but also something *in us solidifies* as we do it. The peculiar significance of action has always been felt in ethics – a significance which goes beyond mere action. If Schopenhauer was right in thinking that man's character and inner constitution would be fixed and immutable once and for all if there were no such thing as self-realization, as getting through to reality, action would have significance *only as action*. But we feel that an important aspect would be lost, that action concerns us also in a very different sense: as the point at which the ego passes into the external world, where our inner destiny is decided. If we look at an action only from the outside, say with a view to its social value, we never see it as what it means to the agent himself – as part of his destiny.

What has been said applies of course only to unique and unrepeatable actions, not to routine actions in which we are only superficially involved. Actions could be arranged according to the depths from which they arise, or by their distance from the ego. We could then say that some actions are completely egoless; in other actions we express ourselves, in some more so and in others less so. There is a point of view from which all actions appear on one and the same level, that of a process in the external world; and there is another point of view from which they seem to belong to different layers, as it were, depending on how close to or how far they are from the ego.

'That is just like him', 'That is typical of him': we can and usually do say this of actions which express a certain personality; but these actions are to be distinguished from those in which an ego shapes itself, in which something breaks through from the depths; the former are *charged with personality*, the latter *personality-creating* actions. 'An action is expressive of a personality' can therefore mean two different things: that it flows from a personality which always presents itself like this, in which case the personality exists independently of the action, or that it is something by which a personality first comes into being.

There are thus *impersonal* actions – like the vast number of things we do in daily life when we go about our business – and *personal* actions, or better, *personality-related* actions; and the latter divide again into two kinds: those that follow from a fixed personality and those that play a role in the making of a personality. The fateful significance of which I spoke above attaches only to the latter kind of action.

11. THE AMBIVALENCE OF THE WILL

We have spoken of willing and counterwilling. But how is it really: can we will and not will the same thing at the same time? Or is there not always some difference between what we will and what we do not will? I will go to Paris; yes – if only it were not so much trouble and bother; this I do not will. If I say A, I must of course say B: there's the rub. For I will A but not B, even though A and B happen to be connected in this world. This makes it look as if I willed and did not will the same thing.

But would it be *conceivable* for me to will and not to will the same thing? Well, this would be asking whether such an assumption makes sense. Preliminary remark: the negation of 'I will A' is that I do not have such a will, and not that I will not-A. This is obscured by ordinary language; for the negation of 'I will go to Paris' is 'I will not go to Paris', which can mean two different things: that I do not have the will to go to Paris or that I have the will not to go to Paris – perhaps I am refusing to accept such a proposal. Now if we say that we will and do not will something, we mean that we feel a movement of the will to do it and another movement not to do it; but we do not mean to say that we feel and at the same time do not feel the one movement of the will – which would be contrary to the law of contradiction and complete nonsense. Thus logic prohibits us from saying that I have and do not have a movement of the will; but it does not prohibit us from saying that I perceive a movement of the will or an impulse in me to do something and at the same time another movement of the will not to do it. These two movements of the will can of course exist side by side, just as two contrary feelings like tenderness and hostility can exist side by side in our emotional lives: the will can therefore be ambivalent. In addition – though this is only a conjecture – underlying such an ambivalent will there is also an ambivalent feeling of attraction and repulsion.

Thus in the sense explained, I can will and not will the same thing. However, we often find ourselves in a different situation. I will something, e.g. to go to Paris. When it is time to act, I find that something in me does not will, that something 'balks' at it, that there is a countercurrent which manifests itself in various symptoms: inexplicable inhibitions, Freudian slips, forgetting, clumsiness, etc. If I try to see myself as I appear to an outside observer, I get the impression that *I behave as if* I did not will it at all. It is the behaviour of a person who claims to care a lot about taking the trip, but who is animated by a secret counterwill. If I were looking at another person exhibiting this behaviour, I should begin to doubt whether he was sincere about his will and say to myself: he only acts if he had that will; in reality he does not will it at all. But what if I myself am that other person? Of course in that case I can no longer say that I am 'acting' or only 'pretending' to will: for *I* must know if I will or not. Well, my self-consciousness says 'I will', but my actions say 'I will not'. Today this situation is usually described by saying: I will to do A, but unconsciously I will not to do A. Thus the conflict between my will and my actions is interpreted as the conflict between a conscious and an unconscious tendency of the will. This raises a new question. I just said that the conflict is *interpreted* in this way. Should I have said that it is expressed in this terminology? The difference is this: when I say that the conflict is *interpreted* in this way, I mean that I *guess* from my behaviour that there is an unconscious will, but it is still open to question whether there really is one; as when I guess from certain deviations in the course of a ship that there was a storm at the time which drove her off course, but that this does not yet establish that there really was a storm at the time. The deviations in the ship's course only *help* me to put forward the hypothesis, which must then be verified independently of these indications. On the other hand, when I say that the conflict is *expressed* in this terminology, I mean that the 'unconscious will' is *nothing but* an abbreviated expression for my strange unruly behaviour; it amounts to nothing more than that behaviour; so it makes no sense to ask whether there really is an unconscious will. As if someone were to say: the clouds are hurrying along, the leaves are whirling through the air, it is whistling and roaring – but is there really a wind? What could he mean? The wind amounts to nothing more than these and similar phenomena; what more is supposed to be behind them? Now if I say: I conclude from my behaviour that there is an unconscious counterwill – in which of the two situations do I find myself: that of the man who notices the disruption of the ship's course and thinks 'There must have been a hurricane', or that of the man who witnesses the uproar of the elements and says: 'What a hurricane!'? Freud holds the former view, whereas Russell and the behaviourists hold the latter. It seems to me that the latter view does not quite do justice to the facts. For if I regard talk of the unconscious will as a convenient shorthand for describing a certain type of behaviour, I can choose this way of speaking wherever I find a certain course of action, and it can no longer be asked whether or not this description is 'legitimate'. But I cannot help thinking that this view is overly simple and has something artificial about it. For all cases in which there *appears* to be a counterwill – even though the agent himself has no inkling of it – are simply given the label 'unconscious will'. But does this square with the fact that human behaviour is enigmatic, ambiguous and hard to understand? For a person may say that he wills A and also will it sincerely, but be prevented from doing it by various unexpected adverse circumstances; he may even make various Freudian slips which he could not, as it were, 'help' making; he may neglect, forget or omit something; or he may fail at the crucial moment because he is tired, did not get enough sleep, or for other reasons beyond his control, so that he cannot be accused of having 'arranged' them. And even though it looks as if he was animated by a secret counterwill, could this not be the result of a series of repeated coincidences? Such possibilities are excluded from the outset if they are labelled 'unconscious will' by definition. What should we say if someone defined a hurricane by saying: 'Wherever a ship deviates from its normal course, I will ascribe this to a hurricane; this is how I will use the word "hurricane"'? We should find that this is arbitrary and does violence to linguistic usage. But this is the position of those psychologists who ascribe everything that looks like resistance to an unconscious will and then explain that this is just another way of describing such behaviour. An unprejudiced observer will first try to make out in a concrete case whether the behaviour is an instance of apparent or real resistance. And there is a good method available for this purpose: it is well known that a person can be led to *acknowledge* a counterwill through psychological treatment, not in the sense that he has fooled us all along and now stops play-acting, but in the sense that a movement of the will he had repressed and made inaccessible now breaks through into his consciousness with all the signs of a soul-shattering experience. So *that* is the true criterion which justifies us in speaking of an unconscious will. As long as we have *only* the behaviour, talk of an unconscious will can be no more than an hypothesis which has yet to prove itself.

It is therefore possible to will and not to will the same thing: either in the loose sense in which one wills A but not B even though one cannot reach A without B; or in the strict sense in which one either feels two opposed movements of the will – two impulses – in oneself or wills A but has an unconscious counterwill.

Can one will the same thing consciously and unconsciously? What is this supposed to mean? To will something unconsciously means: to behave as if one willed it, and ... [The text breaks off at this point.]

12. HOW DOES ONE COME TO A DECISION?

Let us start with a simple example.

Whilst talking, I become conscious of a pin on the floor, or some dust on my sleeve. Without interrupting the conversation I brush away the dust or pick up the pin. I make no express resolve, but the mere perception of the object and the fleeting notion of the act seem of themselves to bring the latter about. Similarly, I sit at table after dinner and find myself from time to time taking nuts or raisins out of the dish and eating them. My dinner properly is over, and in the heat of conversation I am hardly aware of what I do; but the perception of the fruit and the fleeting notion that I may eat it seem fatally to bring the action about.[6]

This will hardly be called an act of will; the things James mentions are done 'automatically', as we say. But the will is involved to the extent that we could also have omitted the action by directing our attention to it. This distinguishes such an automatic action from a reflex – such as shedding tears when some coal dust flies into my eye – on which the will has no influence.

Now take a case where there is something like an initial impetus.

A man says to himself for example: "I must change my shirt", and he has involuntarily taken off his coat, and his fingers are at work in their accustomed manner with his waistcoat-buttons, etc.; or we say: "I must go downstairs", and ere we know it we have risen, walked, and turned the handle of the door; – all through the idea of an end coupled with a series of guiding sensations which successively arise.[7]

Is this first impetus a movement of the will? Or is the action triggered by the mere idea of changing one's shirt (or going downstairs)? However this may be, we reach a *clear* case of willing as soon as we find some resistance along the way. Take the case where I say: 'I must go downstairs': I go automatically to the door and press down the handle – but it does not move! This interrupts the natural flow of the action.

While hitherto, perhaps, I have been thinking of very different matters, now my attention is centred upon the door; I shake it vigorously, sense the tightening of my muscles, and experience exertion against what opposes me. The idea of opening the door stands firmly and clearly before me as an image of my goal. I "will" to open the door.[8]

That is the simplest, clearest and most common case of an 'act of will' proper.

James says: 'Wherever movement follows *unhesitatingly and immediately* the notion of it in the mind, we have ideo-motor action. We are thus aware of nothing between the conception and the execution.'[9] Accordingly, shaking the door when it refuses to open – an action done 'unhesitatingly and immediately' – would have to be counted among ideo-motor actions and *not* among acts of will – which is surely contrary to linguistic usage. To my mind the essential difference between ideo-motor actions and acts of will lies in overcoming *resistance*, paying *attention* as one does it, and having a feeling of *effort*. If in the heat of conversation I take some raisins from the bowl in front of me, hardly noticing what I am doing, there is a lack of resistance, attention and effort; *therefore* it is an ideo-motor action. If I am a shy dinner guest and I first have to overcome my shyness, I shall have willed.

But not all acts of will issue from *decisions*. A decision presupposes some prior reflection or a struggle between conflicting motives, and it comes as the result of an inner conflict. Let us now look at different types of decision-making.

Suppose I come to a decision after an inner conflict. What really happens in such a case? First there is a phase of reflection and deliberation, where we are torn this way and that between conflicting ideas, wishes, movements of the will and motives. We listen to the arguments and then to the counterarguments; we deliberate and decide; it is like a trial in an invisible court of law with two parties facing each other; and in actual fact our reflection often takes the form of dialogue. And when I finally decide, I know why I came to this decision: I can point to all the arguments that led me to it; and it will be difficult to get me to go back on it; and I myself will hardly change such a decision unless new facts turn up necessitating a new trial, just as in the legal case. This process could be called *rational will formation*.

There is the very different case where we have to make a decision that will profoundly affect our lives; now it is not just a matter of reflection, cool judgment and objective finding. We ourselves are somehow involved in the decision. The decision seems to come from a much greater depth, and rational arguments often turn out to be powerless against it. Even when we are tired or asleep, everything in us collaborates in the decision; like a plant it draws its sap from our entire being rather than just our intellect or our ready-made views and values. We feel that such a decision concerns us very deeply, that we ourselves may change in making it. What we have in hand is not just the decision but also, as it were, *ourselves*; this is also why nobody can help or counsel us in such a situation. Nobody in the world can decide for us, just as nobody can live for us. In such a decision we express and realize ourselves, i.e., we help one of the possibilities dormant in us to emerge. We are therefore dealing with a different type of will formation. We wait until the will 'matures' in us. It is no use going over the arguments pro and con over and over again. One day the decision will be there without our knowing how it got there; something will become clear to us and win through, putting an end to wavering, and we shall suddenly know what we will. To repeat, this is not a rational process, even though we finally become convinced that it is the 'right' thing to do. We feel *why* it is the right thing, but cannot quite put it convincingly into words. We can of course always find reasons, but it is fairly certain that we do not do what we will for these reasons. Rather, the will chooses these reasons in order to preserve a semblance of rationality. But is there not also the case where after endless reflection and going back and forth over the same ground I feel a sudden jolt and say 'Now I will'? Maybe so; tired of wavering I give myself a jolt and choose; I should almost like to say: no matter what. Any choice is better than this vacillation. In such a state of indecision small insignificant things may play a part or tip the scales: we may even leave the decision to chance or superstition (counting coat buttons). On the other hand, this kind of decision is very 'loose' and is immediately revoked if things turn out unexpectedly. There is not much 'will' in it.

The following case illustrates a different type of decision: we feel the need to do a certain thing, almost like a voice urging us on, but we evade it, give up, offer a pretext, while knowing in our heart that this is not the end of the matter; in fact the voice returns and does not let up until we finally do the thing – like a prophet who would like to shirk his task but receives a visitation. The thing in question may have been something we considered in our mind but recoiled from – like a suicide who after long hesitation and many false starts does away with himself, or an assassin who finally pulls the trigger, or Macbeth who finally commits his crime. We could speak of preliminary waves of willing.

We find something different where the action is frequently put off or drawn out. In this case the agent may do less than he willed; he may get bogged down in minor details and never get around to doing all he set out to do. The straight line leading to the action gets bent at the half-way mark, and this is where we meet some strangely broken people: those who listen to their inner voice but stop short of doing what it says, the compromisers who never get to do what they really want to do because of all sorts of doubts and scruples, and the half-hearted – all those people who take half-measures, who do in fact do something, but only part of what their inner voice commands. Diametrically opposed to them are people with a tough, determined will: those who stick to a goal for years through thick and thin, take everything upon themselves, move into position and wait patiently for an opportunity to act; when it arises, they seize it with lightning speed and exploit it skilfully, proceeding with the greatest circumspection and taking full advantage of the circumstances. The toughness and persistence of their willing contrasts strangely with the great agility and elasticity they show in action.

Different from all these cases is the sanguine type. Such people have no need of lengthy preparations, inner debates, doubts, or wars of attrition between hostile motives. As soon as they see an attractive and apparently attainable goal, they have already made their decision, apparently with hardly any effort. The resistance to be overcome is greatly reduced in them. The result is quick and easy action, accompanied either by a certain carelessness and the feeling that 'things will turn out all right' or by pure joy at the discharge of energy, or even by a rather contagious enthusiasm. Everyone knows such impulsive, mercurial, optimistic people who are constantly on the move, busy with a thousand plans, bubbling over with life, never discouraged by failure, and instantly ready to launch into some new enterprise. But since their limit of resistance is so low, we get the paradoxical result that despite frequent and effortless willing they do not really get the experience of willing: they are not 'willing beings', but basically ignorant of the will as a slow, calm force.

I knew a man who was, for a long time, unable to make up his mind about a vital question; he put off the decision week after week without being able to sort things out with himself. One day everything was so cheerful – the sky so light, the day so carefree, and he himself so full of new courage that all of a sudden all his doubts and objections vanished into thin air: he himself felt light-headed, as if slightly dizzy, and decided on the action. What happens in such a case is a change in one's entire scale of values, a transposition of a melody from a minor into a major key – the whole world with its feeling tones suddenly changes. The converse also happens: going from a light-hearted to a serious state of mind in which things appear deeper, weightier, more important. Our previous pleasures appear stale and empty to us, and an idea we may have dismissed earlier appears before us in all its splendour; we are filled with a new seriousness. Perhaps we suddenly perceive the tragic accents of life. In undergoing such a change we can bring ourselves to make decisions for which we might never have had the strength otherwise. Much of what we are told about conversion belongs here: the convert looks around him as if he had suddenly woken from a dream, and everything appears new and changed to him. He cannot understand his previous life and is so shaken by the change within him that he sees in it the intervention of a supernatural force.

Incidentally, the change in emphasis is never an isolated phenomenon: as a rule it goes hand in hand with a change in our whole attitude, the goals we set ourselves or approve of, the perspective in which we see things, even the theoretical convictions we form. Now how is it when a new truth, say a religious or political doctrine or conviction, enters our minds: does this depend on our will? It depends not so much on the will as on all the circumstances that make us feel inclined or disinclined to accept it, but we also become aware of an element of willing. Suppose a view does not 'suit' us because it is incompatible with our interests, self-esteem or ambition, or does not fit into our conception of the world. Such a view meets our *resistance*. If we consider it anyway and try to examine it impartially, we use our will. In this sense it must be said that the will runs even through purely theoretical convictions: it takes a certain attitude of will to acquire them and even to stick to them. Of course what is called religious 'belief' always includes a strong component of willing: for to 'believe' something is not simply to hold it to be true intellectually, but rather to accept it of one's own free will, to submit to an authority, or to sacrifice or suppress one's own criticism; and this comes out in the phrase 'to give/attach credence', the formal equivalent of the common verb 'to believe'. Conversely, a conviction can change the kind of willing. A fatalist will submit to his fate; a believer in the possibility of character-building will work at himself; a believer in progress will try to intervene in the course of history, etc. even though the issue is again complicated by the fact that such supposedly theoretical convictions are often merely rationalizations of basic orientations of the will.

Opinions could in principle be divided into two classes: those that change with our 'attitude' or 'scale of values' and those that do not. I suspect that the second class is empty. For the will reaches even into our intellectual life. The opinion that the external world is real rather than a dream has to do not only with our perception and reflection, but also with our volition. Who does not know that we are sometimes overcome by a mood in which everything appears unreal:

> Alas, how all my years have vanished!
> Did I dream my life, or is it real?

If we have no hopes or fears, the world slips away from us: we still see it, but are separated from its reality as by a glass wall; it is still 'there', but it is no longer the old, familiar world; it has lost something we know not what; it is as if a shadow had fallen on the sun. This shows that the attitude of will plays a part in the formation of the concept of reality – or perhaps better, of the *feeling* of reality. The question would be, e.g., whether a being unable to feel anything would have the same conception of reality. There seems to be more to the belief in the reality of the external world than comes out in the usual analyses of academic philosophers.

I am writing a poem. It is as if I followed the dictates of an inner voice; I listen to the words, which rise like small shimmering bubbles, redolent of mystery and unfathomable night. But this is not quite right: I reject ideas, I sift and screen them; and this involves a kind of willing. Admittedly the idea does not come from me; but what I make of it – whether I help it into being in one piece or allow it to be mutilated along the way – that is up to me and *I* am responsible for it. I cannot quite say of a poem that I write it, nor that it writes itself, nor that it is written through me by a mysterious third something. Active, passive and reflexive fail: we would really need a new verb form to express the process.

These examples are only meant to show how deeply the will penetrates thinking and doing, even in cases where it does not look like that at first sight. Let us now return to considering different types of will formation. There is the case of the peaceable or timid person who is always humiliated or made the butt of practical jokes until one day he plucks up courage and does something nobody had thought him capable of, not even he himself, so that he is startled by his own boldness. Afterwards he is not up to doing what he did: he shrinks back from it and becomes small and insignificant. But at the moment he did it, something raised him above himself, so that the person who did it is not really the same as the person who looks at it afterwards with a baffled expression on his face. We could speak here of an unexpected explosion of the will.

Or take the case of a person who wavers for a long time, perhaps because he is still studying the situation, or because the reasons do not yet seem good enough to him. But by some accident he gets going; he takes a small step and now receives all sorts of reinforcements from all sides; his will forms – and his decision firms up – only as he begins to act. This is related to the fact that sometimes an idea occurs to us only as we try to express it. When we start a sentence, we may have only a vague sense of the direction in which it is going until the idea we wanted to express becomes clear to us as we speak. Just as our ideas often become clear only while we are speaking, so the will may form only during the action. In his essay 'On the Gradual Composition of Ideas in Speech' Kleist gives a good example of how *both* an idea and the will can form only as one speaks:

I think that many a great speaker, at the moment of opening his mouth, did not yet know what he was going to say. But the conviction that he could draw the necessary wealth of ideas out of the circumstances and the resulting mental agitation emboldened him to make a start, trusting to his luck. I am thinking of the 'thunderbolt' Mirabeau unleashed on the Master of Ceremonies who, after closing the last monarchic session on June 23, in which the King had ordered the estates to disband, returned to the assembly hall where the estates were still sitting and asked whether they had heard the King's order. 'Yes', replied Mirabeau, 'we have heard the King's order'; I am certain that when he made this mild beginning he was not yet thinking of the bayonets with which he would end his speech. 'Yes, sir', he repeated, 'we have heard it'. It is clear that he did not yet quite know what he wanted. 'But what gives you the right', he continued – and suddenly a source of outrageous ideas opened up to him – 'to give us orders here? We are the representatives of the nation!' That was what he needed – 'The nation gives orders and does not receive them' – to rise at once to the heights of audacity. 'And to make it perfectly clear to you' – and only then did he find something to express the entire resistance he felt in his soul, now up in arms – 'tell your King that we will not leave our seats except at bayonet point.' Whereupon he sat down contented with himself ... Perhaps it was ultimately the slight twitch of an upper lip or the suggestive movement of a cuff which – in this way – caused the overthrow of the established order in France.'

There is another strange phenomenon which Robert Musil describes as follows:

And in the instant when she reached this conclusion it seemed to her that her feelings began to get under way, like the movement of a crowd that had been delaying for a long time: she had been forcing herself to pretend that she had forgotten Ulrich's invitation to visit him, but no sooner had the first sensations begun to detach themselves from the dark mass and move slowly ahead than an irresistible urge to run and shove entered into those further behind. Though she could not make up her mind, it had made itself up of its own accord without taking any notice of her.[10]

We might say: 'It willed in her'.

At the start we spoke of rational will formation. There is also something like irrational will formation. The most common case is when we simply follow a momentary impulse, without inquiring into the reasons for or against it. Let us now look at a special case of irrational action - say a superstitious one. Superstition may be as good a motive as any: a superstitious person does something because he is convinced that not doing it will bring him bad luck; e.g. he will incur the ill will of some evil power. But suppose someone is 'above' his superstition in the sense that he clearly recognizes a certain feeling in him as superstitious and even makes fun of it; still, this need not keep him from following it. How can an enlightened person be so superstitious, we ask scandalized. What happens in such a case is perhaps this: he hears an inner voice whispering to him: Don't do it, or you will suffer. He may struggle against it and suppress the impulse for a while, but he does not have a clear conscience, so to speak: the monitor in him does not give him any peace, and to rid himself of the idea which torments him he gives in to it, seeing that discretion is the better part of valour. And so it happens that the poor man who, on the *intellectual* level, is not at all taken in by the illusion is nonetheless unable to resist it. How many people have a bedtime ritual they must follow conscientiously if they want to sleep in peace even for a minute. No matter how silly or senseless the particular features of the ritual may appear, the victim does not have the power to evade them because any such attempt would end in an almost unbearable anxiety attack. Most people are predisposed to mental disorders. This leads on to the large area of compulsive actions studied in great depth by Freud and his followers.

Finally, we may will something and do it, but with the best will in the world be unable to say why. If we look at it afterwards, we shake our heads and say: I don't know what came over me. This leads on to cases where the willing subject has the impression that the impulse does not come from him, but that something foreign 'wills' inside him. To quote Freud:

> Obsessional neurosis is shown in the patient's being occupied with thoughts in which he is in fact not interested, in his being aware of impulses in himself which appear very strange to him, and in his being led to actions the performance of which gives him no enjoyment, but which it is quite impossible for him to omit.[11]

There are certain ways of acting based on lack of inhibition which differ from neurotic compulsive action as well as from healthy willing.

> Ask half the common drunkards you know why it is that they fall so often a prey to temptation, and they will say that most of the time they cannot tell. It is a sort of vertigo with them. Their nervous centres have become a sluice-gate pathologically unlocked by every passing conception of a bottle and a glass. They do not thirst for the beverage; the taste of it may even appear repugnant; and they perfectly foresee the morrow's remorse. But when they think of the liquor or see it, they find themselves preparing to drink, and do not stop themselves: and more than this they cannot say.[12]

Do these unfortunates will? Not really; rather, they wish for the contrary; but they do not even have the contrary will: it seems that in this case the paths along which impulses usually travel are so well travelled that even a little stimulus is enough to release the action. The 'sluice gates' no longer work with them.

To study the contrary phenomenon, let us look at cases where the inhibition is too strong. The result is that there is no action whatsoever. We all know people who can never make up their minds and wear out both themselves and those around them. Faced with a decision, they know how to balance the reasons for and against until the result is zero. They manage to raise a thousand scruples and difficulties and are admirably quick and clever at finding obstacles everywhere. In their 'helplessness' they turn to someone else, seemingly to get advice; they also listen to him attentively, but then use so much ingenuity to develop so many counterarguments that in the end one cannot but agree with them; but as soon as the balance of decision threatens to tip to the other side, they are able to track down so many new and unsuspected difficulties that it begins to look as if there is only one thing that really matters to them: to find reasons for remaining in a state of indecision. If only they did not have to do anything; if only they could let reason go round in an endless circle of possibilities and counterpossibilities without ever coming down to a decision! In these cases we begin to see how little the intellect is really capable of, how all it can really produce is arguments and counterarguments, because even the simplest decision in life touches on so many things that the sum never works out exactly. The will is here like a river petering out in the desert, or if you like: here too there is a kind of willing (although the subject is not conscious of it): namely, willing to avoid a decision at whatever cost.

There are decisions we make because we are 'resigned to our situation': because we see that there is really no other way out. I would not *like* to do it at all, but decide to do it anyway because I cannot avoid doing it. Here it will be objected: is this really a decision springing from my will? Am I not rather capitulating to necessity? True; and yet the will plays some part in it: I accept the situation *of my own free will* and decide to do the thing even though I do not really like it. It makes an enormous difference whether I accept my lot and bow to necessity, or do it gnashing my teeth, rebelling against it, and secretly cursing my lot, becoming more and more embittered all the while. *In so far* as I do the former, I am still performing an act of will. But if I do the latter, if I perform a given action because I have no other way out and protest out loud or to myself as I do it, I am acting under compulsion.

Further, it is one thing to will something and another to stick to a decision and carry it through. We live alternately through heroic times, as it were, when we make firm decisions, and other times when we are not up to our decisions. The ebb and flow of the internal sea.

There is also something that could be called apparent will. We expend energy and overcome considerable difficulties to do something we do not really will to do, in the sense that we have never made up our minds or *chosen* to do it; but we do it simply because we have done it before, because we have been trained to do it and it has become second nature to us, because it is considered the 'right' thing to do, because we should lose our self-respect if we did not do it, and so on. In these cases there is really no experience of willing. Think of how many things happen in the world – e.g. the outbreak of war – without anybody really willing anything in particular; it comes about because under the circumstances there is really no other alternative. The Russian Revolution of 1917 was not willed or planned; it just *happened*, somewhat to the surprise of those involved. History is thus full of things that look like acts of will but are not, because there is the whole external struggle but no explicit willing. In such cases we really 'will' only in so far as we fail to will the contrary.

A decision like the one just considered will hardly change our life; for its roots do not go deep enough. But it is possible for a decision to arise from our entire being. In such a case we realize one of the possibilities of our being. But there is another possibility: a decision may be the way we first learn of a deep-seated change in ourselves. That is, our life does not change *as a result* of the decision, but the stream of life has already turned; everything was already prepared for the change in direction; without our noticing it, all our interests had changed; and we first become aware of the change in our life in making a decision. The decision is then an indication of a real change in us. How far apart are all those things we call by the same name!

I conclude with a few questions. Willing seems to depend on temperament; does it also depend on character? E.g., does a proud man will differently from a humble one – I do not mean in what he wills, but in the way he wills it? Can we choose an opinion at will? Can we force ourselves to will?

13. MOTIVATION

If we reflect on what determines an action, we run into a peculiar problem. What moves us to do something is called a motive. Now it looks as if a third thing came in between the motive and the action, namely the will. Guided by the motive, the will brings about the action. But how exactly are we to imagine the relationship between the three? Before we decide on the action, we waver between several motives which push us in different and mutually incompatible directions. We now imagine the will to be something that intervenes in this conflict of motives and helps one of them to victory. If we now ask what determines the will, the answer seems to be: the motive. For we generally conceive of a motive as what moves us to do something, and hence, as what sets the will in motion. So the will chooses or determines the motive; but how can it do that if it is determined by the motive? Further, if the will is determined by some other thing, namely the motive, how can the *will* be what moves us? For then the *motive* would have the power to move, and the will would really be quite superfluous. For if the motive determines the action, this must mean that it brings about the action, makes it a reality. But if the motive is sufficient to bring about the action, why do we need the will?

There is no better way to clarify the situation than by means of the following analogy: if we compare the concept of action with the concept of motion in mechanics, what is it that actually corresponds to the force: the will or the motive? The strange thing is that we get both answers. In fact we notice that the ability to elicit the action is ascribed sometimes to the will, sometimes to the motive. And the reason for this lies again in linguistic usage. To the question, Why did you act like that?, we can reply either: Because I willed it (which merely means that I decided on this action of my own free will and did not do it under compulsion); or: Because I let myself be guided by such and such a motive. This peculiar wavering between will and motive becomes even clearer if we ask someone: What moved you to do this? The answer to it could be either: the will, or: the motive. Therefore we must first come to an agreement on what we are to understand by 'will'.

It is true that there is a struggle between motives in which they come before us, alternately weaker and stronger, and make their claims on us, until one of them finally emerges as the winner. But it is not true that there is a peculiar act that intervenes in this battle and puts an end to it, namely the act of will; rather, willing is the process in which one of these motives wins through at the expense of the others, in which one scale rises while the other one falls. We *will* only in so far as this battle tends towards a decision; as long as the different motives keep the scale from tipping, as long as our attention fluctuates back and forth, we do not yet *will*; but the more clearly one motive rises and the stronger it becomes, the more clearly we feel its rising as a movement of the will; which is why in the struggle of motives we sometimes feel this, sometimes that thing rising in us as a movement of the will; and when the battle is finally decided, we just say: we willed. For willing *consists* in the decision: it is not a power which brings about the decision. But in this battle we are not just spectators, as it were, who watch out of idle curiosity how those motives fight for a place and try to displace one another on the inner stage: in a way difficult to describe we ourselves are somehow present in these motives; we identify sometimes with this, sometimes with that motive; we put ourselves into them, give them our own life-blood; the contest between them fills us with displeasure, even pain; and at the moment of decision we experience something like a feeling of effort. Thus we say: I found it *hard* to decide; which points to the work that went into it, but also to the difficulty of renunciation: of letting go of other possibilities in which we had already been indulging in imagination. Is not this the reason why a hard-won decision often leaves us feeling a little sad? Willing, choosing always means renouncing. On the other hand, the decision, once reached, makes us feel strong and self-confident.

Part of the effort of will thus comes from constantly paying close attention, to-ing and fro-ing, comparing and weighing, feeling tension and uncertainty – which by itself is a source of displeasure. For a lot of to-ing and fro-ing tends to wear us down, and our facility for making decisions, our 'will power', seems on the whole to depend on our vitality; when we are in a bad way, tired or ill, even the easiest decision can cost us a superhuman effort; we mull over it listlessly, unable to make up our minds, put things off until tomorrow, let our minds drift; whereas when we are full of vitality our willing comes easily and spontaneously. Once we have overcome our hesitation and made up our minds, the mere fact *that* we are able to make a decision fills us with joy and confidence. 'Better to will something rather than nothing', we tell ourselves and welcome the decision with relief. Only then do we notice how tired and worn out we are after the hesitation, the to-ing and fro-ing before the decision. Another part of the effort is due to our having to tear ourselves away from a motive which had already captured our imagination, our having to give up plans, wishes, hopes and perhaps daydreams which had already taken root in us and now have to be eradicated. The more we have to *overcome* in us, the harder we find it to make a decision – whether we have to sacrifice our inclinations to what we recognize as our duty, or find that what we have decided to do 'goes against the grain' and is not quite compatible with our nature. And even if the conflict is not between inclination and duty, but between one inclination and another, there remains a residue of sadness, of unfulfilled desire. How strange man is! Reason tells him: you cannot have both; decide on the one and let go of the other! But foolish desire would like to have both, and accepts the loss of the other only with wistfulness.

In the struggle between motives a motive is overcome. And this, along with the feeling of effort we experience in the process, is what is characteristic of willing.

Now what is it that really happens in case of a conflict of motives? To say that there is a peculiar fluctuation of attention or, in Schlick's words, 'a more or less rapid shift of ideas, which alternately appear and disappear, as weaker and stronger, clearer and more confused',[13] is probably to oversimplify matters. For in many cases there are no vivid, clear ideas at all: I let my thoughts drift sometimes in this, sometimes in that direction, with all the indeterminacy peculiar to thinking. The processes are shadowy; while I know what I am thinking of, it would be difficult to make out definite ideas constituting the kernel of my thoughts. The conflict is not therefore one in which mental pictures fade or become sharper, except perhaps in very primitive cases. What, then, is it like?

I should like to say: motives flow through us, take us along for a bit and leave us somewhat altered. Depending on whether we incline to this or to that motive, we see things from a very different angle, feel their relative weights shifting, and find that we ourselves are not exactly the same. To let a motive act on us does not mean: to consider it theoretically. It is a fiction to think that we can identify a subject, an observer, who remains entirely untouched while he focuses on one motive after another. Rather, motives are *part* of the observer; they belong to him, and he cannot be clearly distinguished from them. I cannot say: here am I and here are my motives. No, it is deeper than that: when we are strongly swayed by a motive, we are in a somewhat different state, in another 'mental field': our interests run along different lines; our wishes, plans, prospects and expectations combine in different ways; the possibilities yield new configurations; emphases change; perspectives emerge or vanish. Every motive is, as it were, a focal point around which our momentary thoughts, wishes, judgments and inclinations group themselves so as to form a kind of field. To put it paradoxically, the willing subject is a different person depending on what he wills. As a rule, the change is fairly slight, and yet it is like the beginning of a gentle vibrating movement of the possibilities within him. If the vibrations get stronger, he notices that his personality is in danger of losing its unity: he is then confronted with one of those fateful decisions; he feels it is a matter of life and death for him.

This sheds light on the question why we cannot cope so quickly with ourselves when faced with a far-reaching decision. We try out different configurations of wishes and expectations, as it were; we enter in thought into different situations and try to find the one that least disturbs the equilibrium. We are not so much choosing as experimenting with ourselves and waiting until it 'dawns' in us. And we now understand that this is essentially a tentative experience rather than – a purely intellectual judgment.

14. HOW DOES ONE COME UPON A MOTIVE?

We can inquire into what a motive is by asking: what do we remember when we reflect on why we did something? In general it will be thoughts we had before or during the action. I remember, e.g., having said today: 'I can't stand the heat any more, I've got to go into the water.' Well, then this thought was the motive for going swimming. I may have said these words below my breath or out loud, to myself or to other people. But even if I did not think anything of the kind, but merely suffered from the heat and went into the water, I will call the heat the motive for going swimming. Perhaps I imagined going swimming and had a pleasurable sensation as I imagined it, or some such thing. Now we are tempted to think that there must be something common to all these processes which justifies us in saying that we went swimming *because* it was hot. In reality, the motive of an action can look very different: it may be a thought expressed in words, an idea, a feeling, or something mentioned only after the fact; and there is no reason to assume that there must be the same mental state behind all these processes.

So far we have treated motives as known quantities that confront us clearly and distinctly and make claims on us. But this is not quite right – and certainly not in every case. It sometimes happens that there is at first only a tendency driving or pushing us in a certain direction; we feel it or notice it only afterwards, in what we do. If we now ask ourselves why we do it or did it, we are faced with a new situation. The difficulty is how to formulate correctly what corresponds to this tendency and our momentary overall state. It is like the difficulty we are in when we tell a dream: here too we do not quite know how far we actually reproduce the dream, and how far we remove some of its peculiar indeterminacy by putting it into words with definite meanings, and thus, how far we change it by telling it. Similarly in the case of a motive. I feel some pressure mounting in me, some tendency of the will, and while I am conscious of it and search for reasons to explain it, a motive forms in me. This motive is perhaps the meaning I give to the impulse, a meaning involving my entire personality. The motive need not therefore have been the original one existing prior to the action. It thickens, hardens and takes shape, as it were, only after we express it in words. Thus by expressing it, we *do* something to the motive; and this alone shows what difficulties we can run into in asking for a motive.

In a court of law, we are not content with convicting the accused or obtaining a confession from him: we also try to find out what his *motives* were for perpetrating the act. The same act can spring from different motives, and the will depends as a rule on the perpetrator's prior history (upbringing, parents, etc.), his milieu, the circumstances at the time he committed the act, and the motives that drove him to do it. If someone kills a woman in a passion, say out of jealousy, it will be considered differently than if he had killed her in order to rob her of her money. The distinction between 'manslaughter' and 'murder' is based precisely on the kind of motive at work in the perpetrator. Thus even though motives belong to a person's inner world, their presence is an important element in the judicial process.

Here we already run into a difficulty: for are these motives in the same sense? If we say that the perpetrator killed the women *in order to rob her of her money*, the statement of the motive contains the *goal* he was aiming at. Jealousy, on the other hand, does not contain such a goal, but is the emotion that governed the perpetrator, or the urge that drove him to commit the act. Now are these motives in the same sense? Hardly. We might accordingly be tempted to divide motives into two groups: goal-setting ones and those that are not goal-directed. This is supported by linguistic usage, which distinguishes between 'purpose' [*Beweggrund*] and 'drive' [*Triebfeder*]. The perpetrator's intention to rob the woman is called the 'purpose' of the action rather than the 'drive'. On the other hand, his craving for money will be regarded as his drive. The word 'drive' suggests ambition, vanity, self-interest, curiosity, egoism, miserliness, etc. As indicated already by its etymology, 'drive' points to the manifestation of, or part played by, a driving force or, as we say today, something instinctive. This driving force is something that pushes a person and may gain control over him. A drive *need not become conscious*; when we say, e.g., that someone did something out of vanity or a craving for admiration or envy, we do not as a rule mean that he was *conscious* of these drives; in the same way the word 'drive' often designates something veiled, something strangely hidden. A purpose on the other hand is something *one is conscious of*. We do not say e.g. that vanity is the *purpose* of his action. A purpose in this sense is his *reason* for acting, and thus something he probably thought of before or during the act.

The concept of drive borders on that of *impulse* [*Antrieb*], but there is no sharp boundary between them. An impulse seems to be something that impels one from within, but of which one is at the same time *conscious*, such as anger or fury. We will hardly want to say that someone acted in 'unconscious fury'. Moreover, an impulse seems to be something arising from a *momentary* situation, whereas a drive refers to forces in a person's character which are *permanently* effective, such as curiosity, vanity and ambition, which appear not just once but at every opportunity. And yet 'impulse' is often something weaker: cf. doing something 'on impulse' [in German: *aus eigenem Antrieb* = 'on one's own initiative']. To repeat, the distinction is not perfectly clear: Are we to call the jealousy that drove the man to commit murder the *impulse* or the *drive*? The indeterminacy may be connected with the fact that the statement 'He is jealous' can mean two different things: first, he is having a fit of jealousy; secondly, he tends to jealousy – this is one of his basic character traits. Now if a man acts frequently out of jealousy, it will be called the *drive* (or a drive) behind his action; on the other hand, if it is effective now and again, but on the whole only rarely, it will more likely be seen as an impulse. 'Drive' is often used in such a way that it does not really mean a definite motive, but the disposition to act from certain motives; as when ambition is called the drive behind all his actions. Fear may be regarded as the impulse that led a man to act, and fearfulness as the drive behind his actions. The latter is the characteristic trait of a timid person. Cf. the analogous relationship between 'anxiety' [*Angst*] and 'anxiousness' [*Ängstlichkeit*].

Incidentally, you will notice that 'purpose' and 'drive' (or impulse) go together. When the perpetrator killed the woman in order to rob her of her money, this intention was the *purpose*, while the *drive* may have been craving for money – if this was in keeping with his personality. But if he killed her out of jealousy, we would never call this the purpose of his action, but only at best the impulse to it.

In the example just considered the purpose differed from the drive. Is this always so? Suppose a man walks up to another and knocks him down; asked why he did he, he replies: 'In order to revenge myself for the insult I received from him'. This will be called the *purpose*: it is the goal he had in view. But if we say that he acted 'out of revenge', this seems to point again to an impulse. How are we to distinguish them?

I think the intention is the conscious motive or purpose which guides him, but that it derives its mental energy from the passion for revenge, so that revenge can be called both purpose and impulse, though in a somewhat different sense.

Let us now consider the case of murder committed out of jealousy. Here we can say *only*: he did it out of jealousy, and not: 'In order to ...', for there is no verb that would correspond to jealousy in the way that 'to revenge oneself' corresponds to revenge. (For *being* jealous designates a *state*, not an action.) Now what is the reason for this? How can revenge and jealousy behave so differently? Are we to say that it depends on our *language* whether or not we speak of a purpose – i.e. on whether or not it contains a verb to designate the corresponding action? Are we to say: 'If jealousy had a verbal form like revenge, it could also be a purpose?' But suppose someone explained: 'He acted like this in order to satisfy his jealousy' – would this be giving a purpose? Was the goal he sought and thought about prior to the murder to relieve his passion? No; he was *driven* to do it, but he did not think of his subjective state; he did not act like a psychologist trying to cure himself of his passion. When Napoleon reached out for the crown, he may well have done it *out of* ambition; but did he do it *with the intention* of satisfying his ambition? What a muddle! To do something *with the intention* of satisfying one's ambition means: to do it for one's own satisfaction. But it is wrong to think that one performs an action for the sake of the satisfaction it affords one, as if this was the goal; no, one does it – just because one *wills* it. I think the difference we are looking for is this: I can say: 'I will revenge myself' because I want to bring about a certain state – one in which the person who insulted me has been chastened and I have had my 'revenge'. But there is no mental state I could strive to attain in the case of jealousy. What am I supposed to do? I can indeed want to free myself of my jealousy; but this means changing something *in me*, not bringing about a certain state in the external world. There is thus good reason why in the case of 'jealousy' language did not form a verb analogous to 'to revenge oneself'; and this is also the reason why jealousy can never be the purpose of an action.

Now the word 'motive' is more indefinite than 'purpose' or 'drive' or 'impulse' and seems to embrace all three.

Before agreeing with this, we would do well to get an overview of the variety of things we call motives. The question 'Why did you ... [The text breaks off at this point.]

There is no limit to possible answers – they may contain anything: reasons, aims, intentions, hopes, inclinations, desires, cravings, inhibitions, interests, one's situation in life, consideration for others, duty, social position, awareness of one's right, a surge of passion, ideals, moods, an inner voice, physical condition, something irrational, and God knows what else. The menu of things *for which* something may be a motive is no less varied: a decision, an attitude, a behaviour, a way of thinking, a benevolent disposition, a belief, an action, an omission, a plea, an assumption, a thought, a mood, a hostile reaction, antipathy ...

This variety alone must make us sceptical about finding a simple definition of what is called 'motive' or of its relation to what it motivates, i.e. of motivation. In other words, we shall have to doubt that we can find a formula that would bring out what is common to all motives – as if we could say for example: a motive is the goal one works towards; or: a motive is what sets the will in motion. It should be fairly obvious by now that such definitions are futile. The truth is that there is nothing common to be discovered in the various kinds of motives, for they have nothing in common; rather, language has united many different kinds of things, used in the most various ways and related to one another by different overlapping relationships, into a loose group without a sharp boundary. Analysis divides this heterogeneous complex again into subcomplexes with different narrower boundaries and makes us see the similarities and relationships between them. We can now revert to the question we suspended earlier, whether the concept of 'motive' covers the three groups: purposes, drives and impulses. It makes perfectly good sense to speak of the *motive* for a hostile attitude, but not of a purpose for this attitude or of a drive or impulse to it. We can have a motive for believing in God or wishing for the soul to be immortal; but can we have a purpose in doing so or an impulse to do so? This shows that the concept 'motive' extends beyond the other three concepts, so that the relationship between them can be illustrated by the following diagram:

MOTIVES
⏞
...drive...impulse...purpose...

Before going on, let us stop for a moment and ask: For what things are there no motives? Or more precisely, are there things for which it makes no sense to speak of a motive? To be sure, a *motive* does not have a motive; thus it would make no sense to ask what motive I had for deciding on a certain motive. For the *decision* consisted precisely in *one* motive's emerging victorious from a conflict between different motives, and it makes no sense to talk as if there were a new motive behind that decision. For in that case we could go on to ask what was behind *that* motive, i.e. why I let myself be guided by just that motive in choosing among the motives, and so on. Our mistake here is to follow a false analogy, misled by the use of the word 'choice'. We imagine that we choose between different motives the way we can choose, say, between different boxes in front of us. In the latter case I can ask *why* I chose just this box rather than some other, and I will probably be able to offer something or other as motivating my choice. But when I am faced with a decision and alternate irresolutely between different motives, it is not the case that 'I choose' one among those motives and that this choice is itself guided by a motive. The words 'I' and 'choose' are in fact ill suited to describe this state of affairs. It is more like this: I sometimes live in this and sometimes in that motive, become immersed in them, until one of them just proves to be stronger and realizes itself in action. I can choose between the different possibilities before me: I can do either this or that, and when I decide to do the one rather than the other I have a motive for it, and in deciding I am at the same time choosing between possibilities *in* me. But I am *not* choosing *between* motives. That is the sense in which we can say that a motive is not something external in front of me, like a box in the above example.

A *perception* does not have a motive either. I can ask: Why did you *do* it? Why did you *will* it? Why did you *plan* it? Why did you *think* it? But not: Why did you see it? (Unless I am asking about the *cause* of that perception.) In perception we do not *do* anything, we are *receptive*. This invites the question: Are *all* states in which we behave passively motiveless? Certainly not: we may have a perfectly good motive for doing nothing; as when I remain calm in the face of provocation; or when I observe calmly something I could have prevented; but let us turn our attention to other cases. What if I hit upon an idea that leads me, say, to make a discovery – can I ask whether I had a *motive* for it? No; and we also assume that the idea 'came to me' rather than that I 'made it'; so I am not active in this case. And what if I make a joke? The funny idea is not *my doing*; and yet the making of the joke can certainly betray an intention, and to that extent it does again make sense to speak of a tendency that betrays itself in the making of it – say malice or aggression. Whether we want to regard this tendency as a *motive* is a different question. Incidentally, it is easy to see that a further element of willing comes into play; for I can abandon myself to the peculiar state of relaxation in which jokes actually occur; I can try to see, as it were, if a joke occurs to me; or I can 'balk' at a joke about to occur to me. Up to a point it is *up to me* how far I yield to the play of ideas or relax my conscious control. Of course an idea either occurs or does not occur, and it cannot be produced at will; but it is up to me how far I enter the state of mind favourable to the occurrence of ideas. Moreover, *what* occurs to me can also betray a motive, though I may not become conscious of it until after I have made the joke. To that extent a joke is a means of letting unconscious motives emerge.

Suppose I write a poem. There is the poetic inspiration: it either comes or does not come, and I cannot command it. There is a motive for it. There may also be the drive that pushes me to write the poem: in the case of a love poem the motive is obvious. In addition there is an element of willing: I yield willingly to that peculiar relaxation or sink into the mood in which the poem comes. As we have already seen, I cannot really say of a poem that *I* write it, nor that it writes itself, nor that it is written by a mysterious third party through my agency; to describe the origin of a poem we would need a new verbal form that is neither active nor passive nor reflexive.

But even in the case of perception things are not as simple as we first assumed. It is an open question whether there is something like a 'plain perception' where we just see and/or hear something without adding anything of our own. What is certain is that we often read something into a perception, or omit and suppress something, and that these changes are in a sense *our doing*. But above all we *order* perceptions in a certain way, so as to combine them along certain lines. Let me try to clarify what I mean by examples. If we study the paintings of a certain painter, say van Gogh, and look at them often, immerse ourselves in them, or even copy them, we find that we begin to see things in his style: wherever we see a wheat field, flower garden or row of hills, we find these pictorial motifs, these lines of force, imposing themselves upon us and dominating everything we see. I should like to say that the perception is *stylized* in a certain way under the influence of the painting; and if we study the works of another painter for some time, we will see *his* motifs cropping up wherever we look. Let us take another example. If we go into the street immersed in thought and think intensely of a certain person or expect him eagerly, we can be sure to see him even in his absence – in someone else's traits or figure. The expectation makes use of any resemblance to turn into the perception of the longed-for person. And if we do not want to see something, we simply do not see it. Freud observed how children simply deny something that does not fit in with their convictions. From this we can infer that there are organizing forces at work in perception. My knowledge infiltrates perception; but so do wishes, expectations, my style, my convictions. Perception is interspersed and interwoven with these elements. We perceive not only what we perceive: we add, read things into it; but we also omit, order and shape. These are unconscious processes which intervene in perception and change it. Things are not as a primitive school assumes them to be: we do not stand face to face with reality and simply register it like a camera – as patches of colour one next to the other, which are supposed to make up the 'given'. No, we perceive with our entire being, not just with our sense organs. If a child does not perceive something because there is some resistance in him or because he does not want to admit it, we can speak of unconscious motives. Thus perception is not entirely free of unconscious motives.

15. HAVING A MOTIVE AND BEING MOTIVATED

We can ask someone: 'Why on earth do you feel such antipathy towards him?' If he explains it to us, e.g., by offering some ground or other for it, we say, or might say, that he has 'motivated' it. It appears that 'to motivate' [*motivieren*] means as much as 'to give a ground or reason'. In this sense we speak, e.g., of unmotivated aversion, unmotivated hesitation, an unmotivated request, and mean in all these cases that there is no sufficient, valid ground or reason for it. The word 'ground' in one of its meanings derives from legal language and means *legal ground, or ground that justifies something*.

A 'badly motivated request' is a request that is not sufficiently grounded or justified; whereas a 'well-motivated request' is a justified one. We can speak of unmotivated hatred, but not of unmotivated love: love is always unmotivated. There is unmotivated hatred where there is no convincing ground or reason for it (the cause may be something purely physical). The same is true of an unmotivated procedure, an unmotivated dismissal etc. These examples seem to show that 'motivated' means the same as 'grounded' or 'justified'. It is strange that in that case 'motivated' has nothing to do with 'motive': 'How does he motivate what he did?' does not mean: 'What motives does he give for it?' but 'How does he ground it?' or 'On what grounds does he try to justify it?' It will not occur to anyone to speak of the motive of a sad or cheerful mood, but he may on occasion speak of a *ground or reason* for it.

'Unmotivated sadness' is a mood not grounded in any events in the external world; we speak of an unmotivated, as opposed to a motivated, grounded sadness, even though sadness does not spring from a motive. We can say: an unmotivated decision, but not: an ungrounded decision, or: a motiveless decision. We call vanity a drive or spring of human action and ascribe it to someone as a motive. On the other hand, we would not say that his action was *motivated* by vanity, nor that it was *unconsciously* motivated. These examples make it clear that 'motivated' does not mean: arising from a motive.

But this would be too simple a picture. Suppose someone explains his aggressive behaviour by saying: I became heated, and when he made that remark I got carried away and uttered those words. What is he doing? He is making what he did understandable, he is motivating it; but we would not say that he is *grounding* it; for this would mean that he had a good, convincing ground for his utterance, one that justified it; whereas he himself perhaps regrets his words and only tries to make us understand how he came to utter them. In this connection 'to motivate' means: 'to make understandable', 'to explain'.

This brings us to the use of words we find in literary discussions, e.g. when the question comes up whether a playwright has sufficiently 'motivated' a character: we then want to know whether he has made him understandable as a human being, perhaps with the implication that his action only *seems* right. We thus say, e.g., that Schiller 'motivates' Wallenstein's betrayal, i.e. lets us have a glimpse inside him, so that we are no longer repelled by him, but feel human sympathy, understanding, and perhaps even respect or admiration for him. Here 'to motivate' means: to make understandable, to bring closer as a human being, to win sympathy, and thereby to justify at least to some degree.

16. FATHOMING A MOTIVE

There is a curious process that might be called 'fathoming a motive'. We did something, something strange, something out of the ordinary, perhaps something that came as a surprise even to ourselves, or something we did not think we were capable of doing and whose possible consequences we had not seen clearly at the time. Afterwards we ask ourselves what it was that actually made us do it; we would like to make our behaviour understandable to ourselves. At first we are not at all hesitant about naming the purpose or motive. But as we think about it and try to remember exactly what happened, it takes on a different appearance: we see that this motive does not really explain it or only partly, that we have actually been fooling ourselves, that the real roots are deeper, wider and darker. Something else comes out from behind the motive, something that becomes clear to us only now; and this process may repeat itself: a third motive may emerge from behind the second, go even 'deeper' and make our action appear in still another light. Thus it may happen that in the course of our 'soul-searching' our glance shifts from one thing to another, penetrates deeper, and brings to light other hitherto hidden or half-suspected connections. What really happens in such a case?

First, is it like this: do I observe my own action and tell myself that it is quite unlikely that I acted from that motive and now look around for other more satisfactory explanations? Do I *compare* my action with that of people who were in a similar situation, and do I thus indirectly reach the conclusion that the motive I first gave was not the right one after all? No; I do not think of other people who were in a similar situation; perhaps I do not even know such people or what happened inside them; I search *myself*, I try to bare my own soul. For, strictly speaking, two decisions are never the same: each time the circumstances, the antecedents and the agent's personality differ – so what good would it do me to refer to similar cases? Would it make me more certain? But quite apart from this, it is wrong to regard fathoming a motive as a *theoretical, intellectual* process; it is, much rather, focusing on ourselves, not fooling ourselves, listening inwardly, and acknowledging feelings we have never expressed before or been clearly aware of in ourselves. It is *this* rather than constructing a theory about our own behaviour and verifying it afterwards. And this goes with the fact that soul-searching is a very serious matter and almost always attended by signs of emotional upheaval: we suddenly see what drove us to do it; we now remember such and such emotional reactions and gauge their full significance; we now see that the act did not originate at the moment, but had deep roots in our earlier life, and how this motive was somehow already present in our dreams and daydreams or fantasies. Yes, it is both recognizing and acknowledging. I think it makes good sense to say that we are shining a light into the hidden recesses of our own soul. It would of course be best to analyse a real case, but there are understandable objections to it. Who will expose his own most secret feelings to the public? And even if someone had the courage to overcome his scruples, would he be perceptive enough to track down the smallest signs and reactions in himself and to represent them and express them adequately?

Let us observe instead how a writer represents the way one approaches a motive.[14] Raskolnikov goes to see Sonya to tell her who killed her friend Lizaveta: it was he himself. His whole behaviour during this conversation is extremely odd.

'What's wrong with *you*?', Sonya asked in alarm. He could not utter a word. He had imagined it very differently how he would confess 'it' to her, and he did not understand at all what was happening to him ... it became unbearable; he turned his deathly pale face towards her, his lips twitched convulsively and tried in vain to 'express' something. He then started talking in broad hints: he knew who the murderer was, and so on. 'How?', Sonya asked. 'Take a guess', he said with a faint twisted smile ... another dreadful minute passed. They looked at each other. 'So you can't guess it?', he suddenly asked with the sensation of falling from a bell tower. 'N-no', Sonya whispered almost inaudibly. Then a strange thing happened to his gaze: Sonya recoiled from it, and her fright communicated itself to him. 'Did you guess it?', he asked at last. 'Oh God!', a terrible scream escaped from her breast ... for she could not say that she had suspected something like that? And yet, now that he had said it, it seemed to her that she had actually been prepared for *that very thing* ... he had imagined it very differently how he would make his confession: and it had happened 'this way'. Sonya threw herself at his feet. 'A feeling he had not known for a long time flooded his soul like a wave and softened it. He yielded to this feeling; two tear-drops fell from his eyes and got caught in his eyelashes.'

This is how Dostoyevsky describes how Raskolnikov forced himself to confess. This raises the question: Why did he do it? What for? From the subsequent description we get the impression that in talking things out with Sonya he first became clear about his motives; before this he had been much too busy with the external situation, the dangers, the need to cover up the traces, his own unmanageable nerves, and only now does he get a chance to think about it and to give Sonya – and himself – an account of what drove him to do it. It is a striking fact that he does not answer the question all at once, but in a series of steps, even though he does not want to keep any secrets from Sonya and has come for the express purpose of confessing and unburdening his heart. He gives the impression of someone who is himself gradually becoming conscious of his motive during this talk.

Step one: 'How could *you* ... *a person like you* ... how could *you* do it? No, that is not possible!' 'Well yes, in order to rob her!' he said exhausted and at the same time a little annoyed. 'Were you hungry? You ... wanted to help your mother! ... Yes ...?' 'No, Sonya, no!' he stammered, and looked away and lowered his head. 'I was not as hungry as all that ... I did want to help my mother, but that ... that too was not the main thing ...'

Before reaching step two, Raskolnikov goes from ecstasy through despair to feelings of love. 'You know, Sonya,' he said suddenly as if thrown into ecstasy, 'if I had murdered her only because I was hungry, I would now be ... happy'. 'And why do you care, tell me, what good would it do you', he exclaimed soon afterwards as if in despair, 'what is it to you if I confess that I did something evil?' He then asked her whether she would leave him. She cried and embraced him! Again those feelings poured out over his soul like a wave and brought tears to his eyes. 'Sonya, I have an evil heart, don't forget that; it helps to explain a lot.'

Step two: 'And yet,' he said after a while as if waking up, 'and yet it was like this! Yes ... I wanted to be a Napoleon: this is why I murdered her.' He explained that he wanted to test whether he was an exceptional person: one of the few who have the exclusive right to free their conscience of its traditional shackles – if this is what it takes to carry out their ideas. In this sense the great men of history were criminals who did not shy away from spilling blood when it was of some use to them. (He had explained this idea in an earlier conversation with the examining magistrate.)[15] He had thought and thought about the problem whether he was an exceptional person without finding a solution until it occurred to him that there was only one way to settle the question: by putting it to the test. 'Well ... so I stopped thinking about it ... and strangled her'. 'It was just as I told you!' he concluded. But at the same moment he saw through the situation: he suddenly saw that what he had dug up was merely a theory he had been toying with, something he had thought up, which by itself would not have had the power to drive him to murder. The whole thing melted away before his eyes like gossamer.

Step three: 'But all that is sheer nonsense, empty verbiage! Look, you know my mother has almost nothing. My sister is forced to spend her entire life serving as governess. All her hopes rested on me alone.' But his own situation was hopeless. 'Well, and so I decided that once I got hold of the old woman's money I would use it to make it easier for me to get on ... and make myself independent once and for all ... well yes ... that was all ...'. It cost him enormous effort and will power to finish speaking. 'Oh no, no; not that!', Sonya moaned. 'Is that possible? That's not how it was! No, that's not it!' Raskolnikov's reply to this is very strange: 'You see yourself that it was not like this; and yet I told you everything honestly and truthfully.' Yes, that is the peculiar situation in which we find ourselves when we try to fathom a motive: we are perfectly honest, and yet we see that it is not like this, that we have left out something, that what really happened was something different. Raskolnikov now tries to track down the deeper reasons within himself, to penetrate to the truth. He begins with a confession: 'By the way, I am lying, Sonya'. He looked at her in a peculiar way. 'I have been lying for a long time ... none of it was like that – you are quite right. What drove me to do it was something quite different! ... It has been such a long time since I talked to anyone ... I have such a headache ... Exhaustion and helplessness showed through his state of extreme agitation.' It is as if every wave of emotion prepared a new insight and as if he could gain insight only on such a wave. He now considers whether the reason he gave is really valid. He reflects that it would after all have been possible for him to finish university and get a position without getting involved in crime. He compares himself to a friend who is just as needy and is yet struggling through. So why did he kill the old women? Why? He again recalls his situation in great detail. He notices that he had become angry and embittered, retreated into his corner like a spider and sulked at the world. He calls to mind the way he used to live, remembers his mood – 'anger and spite', as he calls it – as well as his thoughts and dreams. He looks at his life from all sides, desperate to find out what gave him the idea of the crime. 'I lay on the couch and thought ... and that is all I ever did ... and I always had such strange, monstrous dreams ... only then did it begin to look to me as if ... No, that's not quite right! Again I am not telling it quite the way it was!' We see how much effort it costs him to bring to the surface something he feels obscurely within himself and cannot yet put into words; how he gets on the wrong track, recognizes his mistake and corrects himself. At last he thinks it was like this:

Step four: At the time he recognized that only those who have the courage to take power will gain power. Those who risk a lot gain a lot. 'I ... I wanted to *risk* it and killed the old woman ... I was just taking a risk ... That's how it was.' This fourth step towards the truth is like the second, except for one difference: earlier he had said that he wanted to prove to himself that he was an exceptional person, a Napoleon, and that like Napoleon he could kill for a higher purpose. He now has a better insight: 'No, nonsense! I did not kill in order to become a future benefactor of mankind! Nonsense! I killed for myself alone, and at the time I gave no thought at all to whether I would become a benefactor of mankind or suck the blood of my fellow men for the rest of my life ... Nor did I kill the old woman because of her money ... I was less concerned with the money than with something else.' Again he falls to brooding. 'Oh, the kind of killing it was! Does one kill like that? I killed myself, not the old woman. At one stroke I killed myself for all eternity!' What set him on this path? What does it all mean? He does not understand it. And when Sonya says that he has renounced God, he thinks:

Step five: He was led astray by the devil. He could not resist the idea of finding out as quickly as possible whether he was a 'louse' like everyone else or a man: someone who could rise above his conscience. But perhaps this idea came from the devil? 'As I lay brooding in the dark, it seemed to me that the devil was tempting me. No, I am not laughing, I am not mocking! he said defensively. I know myself that I was led astray by the devil.' This sounds like a 'lame excuse'; as if he simply wanted to evade his responsibility and shift the guilt, claiming that he had been a victim. But we have no reason to distrust his words. There is nothing in the context of the conversation to suggest that he wants to clear himself, embellish what he did, or pretend to be better than he was. On the contrary: he makes every effort to exclude the contribution of the 'nobler' motives and to blame only himself. 'Nonsense, empty verbiage!' With these words he dismisses any interpretations. But he would like to understand himself and to clarify in conversation what happened within him, and he thus comes to announce in the end that it was the devil who led him on and proved to him, if only after the fact, that he was a louse like everybody else. I think he is right in that he really felt something come over him at the time: a dark and evil temptation – he finds it 'very odd' that he always had 'such monstrous dreams'; he says of himself that 'the idea came to him'; he describes it as something foreign which does not originate from his own will and takes possession of him. These thoughts, dreams or whatever it was appear to him demonic – inspired by the devil, he thinks.

We are thus shown a person who *feels his way* towards a motive. It is not as if he knew clearly *why* he committed the crime; for he tries to understand it himself and hesitates in judging his motives. Fathoming a motive is a painful soul-stirring process. What does it consist in? He looks back: he remembers the thoughts he had before the action, his strange musings; he calls to mind his sombre, embittered mood, his fantasies and dreams; but that is not all: he considers his whole way of life, compares himself to other people and asks himself how he would have acted in their place; but he also asks himself what he himself would have done under such and such circumstances – he conducts thought experiments, as it were, without of course being entirely certain of the results; and all this, to repeat, not in a theoretical frame of mind, but in the greatest excitement and on the brink of exhaustion. *This* is where he looks for information about himself and his character. Incidentally, we learn even more about him – not in his discussion with Sonya, but earlier, in the narrative itself – where we are given a very important clue: the letter he receives from his mother. From it appears that his sister is ready to 'sacrifice' herself and to marry a man she does not love, even a man she despises, in order to secure *his* future. This adds a new twist to the story: yesterday's dream, an 'intoxicating, seductive, mysterious spell', suddenly outgrows the planning stage, moves menacingly closer, and changes into something entirely different: he understands that he must *really* act, he has no time to lose, he must carry out his plan – or else watch his sister turn into another Sonya.

17. THE EXISTENCE OF MOTIVES

The last few remarks should give us pause. If a motive were simply present like the blue of the sky I am now gazing at, how could I be mistaken about it? Am I ever deceived about the colour of the sky? But we only need to think of Raskolnikov to see how easy it is to fail to sort out the innermost motives of one's actions in spite of a passion for sincerity. But this is not the only thing that should arouse our suspicion. Motives are actually unstable and retreat, as it were, before criticism. Let us take a very simple example. I would like to go to the theatre today. Why? Because I want to see the play; because I am interested in X's acting; because I am free this evening and do not feel like spending it alone; because I am a little tired and hope the diversion will do me good; because I suddenly feel like going out. I can give different reasons. But strange to say, if I like I can raise doubts about each of these reasons. Is it really the *play* that is attracting me? Is actor X so important to me? Am I really so badly in need of diversion? Can I be so certain of all this that there is no room for doubt? And lo and behold, as soon as I begin to doubt and turn a reason this way and that, eye it suspiciously from all sides and look out for other possible explanations, it retreats, yields to criticism, refuses to stand up. The situation is very different in case of a solid perception or pain: there it is, and no matter how often I turn it this way and that and look at it from all sides, I am unable to change anything about it. But a motive is intangible like a cloud. Even if I believe, as I do in the case of the theatre visit, that I know my motives fairly well, a more searching analysis shows that I am far from certain about the matter, or at least that I cannot exclude with certainty the contribution of other, apparently remote factors. We are all familiar with the fact that when we make a vital decision and think we have a convincing reason for it and then think about the situation years later, we perceive the contribution of a whole series of other factors which we missed earlier: which is why life often looks so very different in retrospect; and the deeper we delve into the whole process, and the more severely we take ourselves to task, the more quickly we lose faith in the fact that we really did it from that motive, or from that motive *alone*. In short, a motive is not a hard and fast fact with no ifs and buts about it. What, then, is it?

Before trying to answer this question, we shall have to ask: How is it possible to be uncertain and even deceived about one's own motives? One common expedient is to talk about an unconscious 'resistance', which is supposed to prevent us from looking into our own insides, or at least, to distort or deflect our gaze. But this assumption does not explain the peculiar instability of a motive in very simple cases (theatre visit), nor the fact that – even in the case of a simple action – several different kinds of motives seem to mix or to merge into one. It may well be that some mental 'resistance' comes into play under certain circumstances, but it is surely going too far to appeal to it to explain the uncertainty of *all* introspection. It is hard to believe that there is some force *continually* at work in us barring our view into our own insides, or that motives are entities existing in us in isolation, as it were, ready for action, but merely hidden from us by a more or less laborious procedure called 'censorship' or God knows what. Rather, we shall have to face a much more radical question: *are there any motives at all*?

We say: I did it from such and such a motive, which makes it appear as if a motive was something that existed before the act as the shadowy pattern of the act in our mind, and as if action translated this pattern into reality. But what is it that really happens in action? There are, to begin with, certain driving forces or tendencies called drives; there is, further, a certain tension or excitation, either stored or elicited by a perception, communication, etc.; there is a certain set of dispositions, innate or acquired, called a person's nature or character; in addition there are mechanisms (education or tradition) conducting the tension along certain lines; and perhaps there are also certain objectives, wishes, thoughts and fantasies prior to the action; well, and *all this flows over into action*; it is not as if there was, on the one hand, one particular motive or group of such motives and, on the other hand, the action, and that these motives produced just this action. But we simply *do* something, and later, when we think about it, we *interpret* our action in terms of particular motives. By this I do not mean that motives are pure *inventions*. No; we perceive certain currents or tendencies in ourselves, pick out one of them and call it the motive. But when we later think back to that time, we also remember somewhat different currents, interests, thoughts, wishes, moods and fantasies each time, so we can also connect the action with other currents, and we express this by saying: we also did it from such and such motives. This makes it look as if one and the same action issued from a variety of different motives (Freud's 'overdetermination'), whereas we are only placing the action into different contexts; by connecting it sometimes with these and sometimes with those currents, we draw, as it were, certain lines through the stream of life, all of them intersecting at a certain point; and each such line, when followed up in thought, is called a motive. The variety of motives would accordingly be nothing but the various possibilities of interpreting a given action, i.e. seeing it as the result of a series of currents, wishes, hopes, expectations, etc.

In other words, the view I am opposing is that each motive is an independent piece of reality which must simply be accepted, and that all it takes to produce a given action is a number of such motives working together. I think there is little doubt that we are here succumbing to a mythology. To repeat, we can see an action in different contexts; we then *interpret* it in different ways. But to say that all these motives *were present in reality* seems to me a mistake. It shows a naive conception, probably inspired by language, which assumes without looking that there must be a certain entity behind the *word* 'motive'; it is primitive in the same way as the conception that warmth must be some invisible fluid, or that magnetism is a mysterious something that sits somewhere in space and, as it were, just waits for a bit of iron to stray into its territory so it can pull at it. It could perhaps be said that the mere use of a substantive tempts us to believe in a substance. Remember the strange things philosophers have been tempted to say because of the existence of the word 'time'! Something like this temptation is also present in the case of a motive. Although motives are questionable constructions, we are only too inclined to regard them as entities having a life of their own and to imagine that if we acted that way it was *because* such and such motives were at work in us. And then we are surprised that we can be uncertain or mistaken in giving the motive and we talk of some 'resistance' standing in the way of self-observation.

I should like to draw a parallel to the situation in psychology: in quantum physics, too, it looked at first as if an electron had 'in reality' a definite location and a definite momentum at a given time, but that we were prevented from measuring both at the same time because the tiresome indeterminacy relation frustrated any such attempt. On more mature reflection people had to tell themselves: if it is *in principle* impossible to determine both the location and the momentum of an electron, this means that the concepts 'location' and 'momentum' are *not jointly applicable* to an electron, or that an electron simply does not behave like a tiny billiard ball. And as soon as people had freed themselves of the picture of a particle, the way was clear for constructing a theory according to which it is *logically* impossible to make precise statements about the location and momentum of an electron. If we want to construct a picture of an electron in wave mechanics, we must produce a sharply localized, i.e. very small parcel of waves; but according to the conception of wave mechanics this is possible only if we let material waves, which fill all of space and have *different* wave lengths, overlap one another; the smaller we want the parcel to be, the more waves of different wave lengths we have to use, which means that the velocity with which the parcel of waves expands becomes quite indeterminate. The indeterminacy relation is not a curtain which veils the finer processes from us; it merely expresses the fact that we try to describe atomic processes by means of a picture which just cannot be used.

I should now like to propose that we apply a similar idea to motives. Since it appears that in searching for a motive we never get down to incontrovertible facts of consciousness, since motives are unstable and melt away on a critical view, it is better from the outset not to conceive of them as existing things, but to observe what really happens when we act and then judge our own action.

18. EMOTION AND ACTION

But let us first see if it is true that we can never be certain of a motive. Everyone will admit that it is easiest to get hold of a motive when we are dealing with very simple, unimportant, everyday actions; I put out the light; I am asked why; and I answer: because I want to go to sleep. So here I *know* the motive.

But are there not also actions of some importance done from *one* over-strong motive? Certainly; let us look at an example of an action done in an emotional, passionate state.

There was a time when crimes of jealousy were an everyday occurrence. Take a woman who waylays her rival and pours vitriol in her face. We say: her motive was jealousy, or hatred arising from jealousy. We imagine this hatred smouldering in the spurned woman's soul until it finally flares up in the attack. There is certainly some truth in this; but what seems to me false is the view that *the motive could be clearly separated from the action*, as if the one belonged to the inner world and the other to the world of external events. Looked at soberly, the process is more like this: there are certain currents of feeling which include certain bodily manifestations from the start; this is followed by rising excitation; a certain threshold of inhibition is crossed, and the excitation passes into action. But nowhere in this whole process can we discover a point where the motive ends and the action begins. Just like the facial expression for example, the action is part of what is called 'hatred'. If we tried in thought to subtract from the hatred all that was bodily symptoms or incipient action, the remainder would no longer be hatred, but perhaps a cool and calm thought devoid of all passion. What makes hatred an emotion is precisely the presence of different bodily sensations and manifestations which merge without a sharp boundary into the action; so that the action cannot be clearly separated from the hatred (or jealousy). Call to mind the different degrees or stages of hatred there are, and how each degree or stage corresponds to a certain state of mind and body and a certain form of behaviour. If we feel only a very low degree of hatred or just a trace of animosity towards a person, we betray it by being embarrassed by his presence: we do not smile at him, avoid looking into his eyes, find fault with his remarks or criticize and disparage them without saying anything, show an inclination not to be very pleasant to him, treat him ironically, etc. If the feeling reaches a higher degree, we display a kind of behaviour that might be called 'irritable': we deliberately avoid him or adopt a sharp, hurtful tone in speaking to him; we are nervous and irritated, for 'he gets on our nerves'. We have nothing good to say about him and perhaps express our displeasure at him before others. In more serious cases we may go so far as to speak ill of him, stir up opinion against him, even to the point of making accusations against him; otherwise we do not think about him or try to drive away such thoughts because of their unpleasant associations. At an even higher degree of animosity we hardly miss an opportunity to hurt him – provided, of course, that we take no risks ourselves. The stronger the hatred grows, the more we push aside rational restraints – we 'lose all sense of proportion', as we say; we use any means to try to bring ruin upon him, even if we ourselves suffer in consequence; for it is 'all the same' to us: we are ready to jeopardize our own well-being if only we can do something to him: in a word, our emotional life becomes turbulent; we get excited at the mere mention of his name; we 'change': we are no longer like our former selves; and this goes with the fact that when we reach a high degree we can no longer tear ourselves away from thinking of him; his picture constantly follows us around like a mocking face; we spend our days and nights hatching dark plans, indulge in all sorts of fantasies, etc. From here it is only a short step to translating these plans into action: we rise above all scruples and seek to annihilate the hated person; or the sight of the enemy creates such a surge of passion in us that it passes directly into hostile action, e.g. violence. The reader will notice how inseparably physical and mental processes are intertwined. The various manifestations of hatred are just part of its nature: hatred would not be hatred, or a hatred of such and such intensity, if it did not manifest itself in these feelings, ideas, words, tendencies and forms of behaviour; and, what is most important, the manifestations of hatred merge without a sharp boundary into the action. The emotion is *discharged* into action, and the best thing to do is *to count* the actions *as part of the emotion*. Take the woman who pours vitriol over her rival: what is it supposed to mean to say: she does it *because* she does that to her? The relationship is best described without using the little word 'because': the intensity of her hatred *consists in* her doing that to her. What amounts to hatred is not so much a motive confined to the mind and influencing bodily behaviour from inside; rather, it is a whole which must be *divided* to obtain the action. The picture of a motive as a driving force behind the process is misleading here. It would be better to say: *the hatred flows over into the action* because the simile of flowing makes it immediately obvious that there is no sharp boundary here.

To sum up the results so far, traditional language falsely suggests splitting a state of affairs into an action on the one hand and a motive on the other. An action, it is thought, is visible and public, whereas a motive is inferred from it, invisible and private; but in reality an excitation *runs* into an action, manifests itself in it, infiltrates it, or whatever words you want to use to bring out their interconnectedness. Now is it a good idea to draw a sharp contrast between motive and action?

The last remarks point again in a direction where giving a motive touches on giving a causal explanation. It is characteristic of a causal explanation that it follows up subtle connections between events. By conceiving of the above example in such a way that a certain excitation flows over into an action, we have already employed the language of causal explanation. In this case asking for the motive is much more like asking for the *cause*. What is needed in order to make the causal explanation more complete is to go into the finer details of the process including the various nerve mechanisms. We thus see here how the concept of a motive merges into the concept of a cause. The closer a motive is to a cause, the easier it is to recognize it from the outside and to subsume it under laws. The farther away it is from a cause, the more we have to rely on self-observation. And this brings us back to the question in what sense we can speak at all of the existence of certain motives.

19. MOTIVE AS INTERPRETATION

Our observations have shaken the belief that certain motives are present in us like so many distinguishable entities which pull us in a certain direction and, if strong enough, provoke an action. We have pointed out that statements of motive are uncertain and quick to yield to criticism, and all this suggested that a motive is a kind of *interpretation* we put on our action – an interpretation which is not entirely arbitrary, but still very much dependent on our way of 'seeing'.

Let us now put this view to the test by analysing an example in more detail. I will choose a personal example in this case, even though this means disregarding the reasons I gave earlier against choosing a personal example.

Why did I really undertake to write this treatise on the will? There are all sorts of things to be said: first, because the question interests me; further, because I think philosophers have unduly neglected it; moreover, because I am dissatisfied with what I have read about the will and feel a need for greater clarity; and because I hope that in writing down my ideas I will arrive at some new point of view or other; coming to think of it, the problem of the will seems to me a good case for testing the power of linguistic analysis; besides, I admit I find it stimulating to try my hand at a problem so many important philosophers have bypassed and thus to measure myself against them; yes, I also find the task of writing itself tempting; finally, I had something of a feeling of *tua res agitur*, for I suffer from a certain lack of will power and would like to get clear about the nature of my condition. Yes, all this must have played a part in my decision to write a treatise on the will. But now I must already correct myself: actually I never made a straightforward decision; one day I started writing about this subject, more by way of experiment, and without any serious intention of getting involved in a long-term project; then, as I was writing, I got caught up in the subject, and more and more new questions came up and tempted me to continue what I had started; and some time later I began to think that what I was writing could perhaps yield a finished whole and I continued the work partly towards that end. Some time during the work I must have dropped the idea of curing myself of my paralysis of the will by doing such a study.

So there would seem to be quite a few motives; some of them joined in only later, while others were perhaps helpful in the beginning, but dropped out afterwards. But what is the list I just gave based on: did I perhaps recite the individual wishes, expectations, hopes, etc., to myself before starting work? Not as far as I know; one day I just started to write down ideas which had already occupied me for a long time; I was not conscious of some definite reason why I wrote down those ideas unless it was a kind of curiosity to see where they would lead me. Now that I look back on the whole thing I am indeed inclined to say that I took an interest or found pleasure in these ideas and perhaps promised myself a discovery in the course of working them out. But I notice that this is in fact already an *interpretation* I am putting on the work I started at the time; and I notice that I am by no means certain of this interpretation. Did I really expect to hit upon something new? Well, I dare not say anything definite about this. I imagine that I approached the subject with such an unexpressed expectation; but I could not say so with certainty; the other thoughts – e.g. that this problem has been unduly neglected – also came to me only later, perhaps when I considered whether it was worthwhile continuing the work. But if I simply started to write and felt no trace of any of these motives, what right did I have to pass them off as *motives*? I would say: because I *know* myself fairly well, which means: because I know from experience which motives I can ascribe to myself and which ones I cannot. My list of motives was not therefore based on direct observation; rather, I *assumed* those motives, and I assumed them on the basis of tendencies, inclinations, interests, ways of thinking which I know in my case and which have made their presence known to me at various times in my life. Thus when I say that I *interpret* my behaviour by assuming such motives, I do not mean that those motives are pure fictions. They are based on something real, i.e. that which led me to assume those motives. But the fact remains that they are *interpretations* I put on my action after the fact – interpretations *based on my character* in so far as I know it.

While glancing over what I have written, I notice that I have gradually related my work to different aspects of my nature: my interests, the kind of questions that occupy me, the goals I set myself, a certain kind of ambition which I cannot deny, and, further, my concern over certain impediments to willing which had caught my attention, and so on. I am thus talking, precisely speaking, not of observations but of *interpretations*: such an interpretation seems to make my action understandable and meaningful. I could just as well say that a motive is a kind of *meaning*.

If a motive is an interpretation, I understand at once why I am always so strangely uncertain of it, why further soul-searching can shake my belief in this or that motive or let another motive emerge behind it; and why in all this I have the feeling of something unfinished, always just out of reach. I look at my action through different currents or layers of my ego as through media with different powers of refraction; and each time it takes on a different appearance. The deeper I penetrate into myself, the more aspects of my nature my action seems to reveal – the more 'motives' my action seems to grow out of. This makes it possible to understand why any psychological explanation is ambiguous, cryptic and open-ended, for we ourselves are many-layered, contradictory and incomplete beings, and this complicated structure, which fades away into indeterminacy, is passed on to all our actions.

This view coincides with one of Nietzsche's views.

> The historian need not concern himself with events which have actually happened, but only those which are supposed to have happened; for none but the latter have produced an effect. The same remark applies to the imaginary heroes. His theme – this so-called world history – what is it but opinions on imaginary actions and their imaginary motives, which in their turn give rise to opinions and actions the reality of which, however, is at once evaporated, and is only effective as vapour – a continual generating and impregnating of phantoms above the dense mists of unfathomable reality. All historians record things which have never existed, except in imagination.[16]

The view we have arrived at does at least one thing: it makes it possible to understand how there can be errors about our own motives. If a motive were something whose presence we could perceive directly in us without any effort, if it were an *experience* for example, the occurrence of mistakes would be extremely disconcerting. Not so if it is a kind of interpretation. Since we do not know ourselves completely, e.g. we do not see our own vanity – out of vanity – as clearly as that of other people, errors about our own motives become understandable. There is no longer any need to assume the presence of a mysterious power that keeps us from looking into ourselves; rather, the truth is that it does not even occur to us to interpret an action in a way that does not do us credit. We do not hide our motives from ourselves, but in many cases we are just not impartial enough to acknowledge them.

Further, this view seems to be confirmed by linguistic usage. It is or ought to be surprising that the German word *Motif* is used in what seem to be two totally different senses: that of a pictorial (or musical) 'motif' and that of the 'motive' of a human action. But perhaps we should not be surprised at this? For 'to fathom a motive' must mean: to see the action in its natural surroundings, to embed it in a structure of ideas (clearly expressed or only half formulated), wishes, tendencies, fantasies, dreams, impulses, inclinations, interests, etc. (cf. the example of Raskolnikov). To be sure, such groupings combine to form characteristic, constantly recurring configurations, to which we give the names of jealousy, hatred, vanity, curiosity, thirst for knowledge, love of adventure, etc. But this is not so very different from what happens if we group the visual field in a certain way, bring out certain recurring features, the so-called 'motifs', and strongly emphasize them. As a painter picks out certain motifs from a landscape, so we become aware of a person's motives by grouping his action with many features in his life and, as it were, discovering characteristic figures in it. There is only one difference: wheras a painter can always discover new 'motifs', something seems to set a limit to the discovery of human motives; be it that there is only a limited number of typical configurations in the experiental field, or that language has too limited a vocabulary for them. However this may be, we begin to see the deep connection language hints at when it offers us the same word for the two uses.

It is so easy to say 'Know yourself', as if all you had to do was to turn the spotlight inside. It is no use going into seclusion and becoming immersed in yourself. It is only in contact with other people that we find out what we are; and there we show ourselves in a different light each time. It is not as if we had to act in a different light with each person. As the moon creates a tidal wave, so each person pulls out a different side of our nature. That is the truth in Proust's descriptions: he never tells us what a person is, but merely describes his reflection in other people: in those who look up to him socially or else look down on him; in those who dislike him, those who love him, etc. It is true that we often see ourselves like a retouched photograph, but not always: people like the ones we find in Ibsen are taken with their sinfulness; some even wallow in their depravity and torture themselves like Saint Augustine or like Tolstoy in his later years; and many shudder with pleasure at the thought that they could be terribly evil if they only wanted to. There are also intellectual fashions in this field: in Schiller's and Rousseau's time people felt uplifted, touched and sentimental; in Byron's time they were unhappy in a noble and mysterious way; following Baudelaire they suddenly discovered every wickedness in themselves (every superior person was now the image of the devil), whereas since Nietzsche blond and dark-haired beasts have been thrashing about inside them, not to mention all those things they have discovered in themselves since Freud.

Another consideration points in the same direction. 'Man acts always out of self-love', says La Rochefoucauld; 'no, because of his striving for power', contradicts Nietzsche; 'from feelings of inferiority', say others; 'because he strives for pleasure', 'out of vanity'; we could go on: 'out of ambition', 'out of a love of adventure', 'for the sake of self-realization';

'in order to forget his self and to submerge it in a larger whole'. And oddly enough, there is something 'in' each of these explanations: life can really be looked at in such a way that sometimes this and sometimes that opinion is the right one. And it is really impossible to see how one could decide between them – a rather serious admission. But does this not show the ambiguity of the mental, which can be subsumed under several categories and hence not under a definite one?

The more one looks at all this, the more one is forced to doubt whether there is anything like objective self-knowledge. The conception of one's own ego fluctuates in the course of time; different 'motifs' stand out at different times – and become the 'motives' one ascribes to one's actions. Is there, then, only subjectivism, only interpretation, varying over time and with the period costume of the soul? But that does not go very well with an example like that of Raskolnikov, where we are dealing with *more* than interpretation. What Dostoyevsky describes certainly contains a truth: there really is something like digging down to deeper layers, becoming more truthful, struggling passionately, while things become clearer and clearer. There is undoubtedly such a process of plumbing the depths in which one penetrates to one's innermost motives. So things are not entirely subjective: there *is* truth after all. And yet! When we want to put our finger on it, it will not stand up; when we look more closely at it, it looks different again. It is an interpretation and yet something more than an interpretation, knowledge and yet not quite knowledge: what are we dealing with? I think we should need a concept that would be a mean between three things: knowing, acknowledging and interpreting. For fathoming a motive somehow touches on all three. The difficulty of psychology is precisely that our ordinary concepts are too rigid; we need something looser, more indefinite. This brings out the fundamental character of the mental: everything is equivocal, indefinite, floating. In order to describe the mental we need a *language* that is just as flexible; which, of course, runs counter to our usual way of thinking.

This is the deeper reason why we cannot speak of the 'knowledge' of a motive with the same certainty with which we can speak of the knowledge of an object in the sensible world, and why the conception of a motive as interpretation is not quite satisfactory. Getting at a motive is a process *sui generis*; it cannot be correctly described in terms of 'acknowledging', 'knowing' or 'interpreting'. To put it paradoxically: motives are things which are never perfectly real and never perfectly unreal.

NOTES

[1] Robert Musil, *The Man without Qualities*, translated by Eithne Wilkins and Ernst Kaiser, London: Secker & Warburg, 1953, I: 128–9.
[2] Arthur Schopenhauer, *Essay on the Freedom of the Will*, translated by Konstantin Kolenda, New York: Liberal Arts Press, 1960, p. 62.
[3] Friedrich Nietzsche, *Beyond Good and Evil* (*Complete Works* XII), translated by Helen Zimmern, New York: Macmillan, 1907, sec. 109.
[4] Nietzsche, sec. 107.
[5] Schopenhauer, p. 45
[6] William James, *The Principles of Psychology*, New York: Dover, 1950, II: 522–3.
[7] James, II: 519–20.
[8] Moritz Schlick, *Problems of Ethics*, translated by David Rynin, New York: Dover, 1962, p. 33–34.
[9] James, II: 522.
[10] Musil, II: 393.
[11] Sigmund Freud, *Introductory Lectures on Psychoanalysis* (*Standard Edition* XVI), translated under the supervision of James Strachey, London: Hogarth Press, 1963, p. 258.
[12] James, II: 541.
[13] Schlick, p. 33.
[14] Cf. Fyodor Dostoyevsky, *Crime and Punishment*, part V, ch. 4–5.
[15] Cf. Dostoyevsky, part IV, ch. 5.
[16] Friedrich Nietzsche, *The Dawn of Day* (*Complete Works* IX), translated by J.M. Kennedy, New York: Macmillan, 1911, sec. 307.

INDEX OF NAMES

Alexander the Great, 61
Angelus Silesius, ix
Aristotle, 4
Augustine, St., 135

Baudelaire, C., 135
Bayle, P., 36
Berlin, I., ix
Borel, E., 40
Byron, G.G., Lord, 135

Christianity, 4, 27, 43–4, 49
Cicero, 3

Darwin, C., 35, 43
Dostoyevsky, F., 36, 121–5, 134, 136–7

Ecclesiastes, 9–11
Epicurus, 4, 44

Freud, S., 71, 95, 96, 105, 118, 135, 137

Genghis Khan, 41
Goethe, J.W. von, 43
Gogh, V. van, 117
Grassl, W., xv, 35

Hampshire, S., ix
Hildebrand, D. von, 39

Ibsen, H., 135

James, W., 97–8, 105, 137
Job, Book of, 28

Kafka, F., 70
Kant, I., vii, 5–6, 36
Kierkegaard, S., 36
Kleist, H. von, 103–4
Kraus, O., 39, 52

La Rochefoucauld, F. de, 135

Marxism, viii

Menger, K., ix, 52
Mirabeau, H.G. Riqueti, Comte de, 103–4
Musil, R., 69, 104, 137

Napoleon, 60, 89, 114, 122–4
Neurath, O., vii–viii
Nietzsche, F., 35, 43–5, 47–9, 89, 134–5, 137

Pascal, B., 36
Plato, 4–5, 27, 32, 39
Proust, M., 135

Quinton, A.M., ix

Rousseau, J.J., 135
Russell, B., 40, 96
Ryle, G., x, xiv

Schächter J., vii–ix, 7–32
Scheler, M., 39, 51–2
Schiller, F. von, 135
Schlick, M., vii–ix, 1–6, 21–2, 52, 98, 137
Schopenhauer, A., 12–13, 17, 36, 50–2, 83, 89–90, 93, 137
Shakespeare, W., 61
Socrates, ix, 3–4, 27, 29, 31–2, 36
Sophists, 36
Sophocles, 9, 11–12
Spinoza, B., 36
Stoics, 4, 42, 44
Strindberg, A., 35

Tolstoy, L., 35, 135

Velde, G.M.H. v.d., vii
Vienna Circle, vii, ix, x, 21

Waismann, F., vii, ix–xv, 33–137
Weierstrass, K., 40
Wittgenstein, L., vii–viii, ix–x, 52

Vienna Circle Collection

1. Otto Neurath: *Empiricism and Sociology*. With a Selection of Biographical and Autobiographical Sketches. Translated from German by Paul Foulkes and Marie Neurath. Edited by M. Neurath and R.S. Cohen. 1973
   ISBN 90-277-0258-6; Pb 90-277-0259-4

2. Josef Schächter: *Prolegomena to a Critical Grammar*. Translated from German by Paul Foulkes. With a Foreword by J.F. Staal and the Introduction to the Original Edition by M. Schlick. 1973   ISBN 90-277-0296-9; Pb 90-277-0301-9

3. Ernst Mach: *Knowledge and Error. Sketches on the Psychology of Enquiry*. Translated from German by Thomas J. McCormack and Paul Foulkes. With an Introduction by Erwin N. Hiebert. 1976 ISBN 90-277-0281-0; Pb 90-277-0282-9

4. Hans Reichenbach: *Selected Writings, 1909-1953*. With a Selection of Biographical and Autobiographical Sketches. Translated from German by Elizabeth Hughes Schneewind and others. Edited by M. Reichenbach and R.S. Cohen. 1978, 2 vols.   Set ISBN 90-277-0892-4; Pb 90-277-0893-2

5. Ludwig Boltzmann: *Theoretical Physics and Philosophical Problems. Selected Writings*. Translated by Paul Foulkes. Edited by B. McGuinness. With a Foreword by S.R. de Groot. 1974   ISBN 90-277-0249-7; Pb 90-277-0250-0

6. Karl Menger: *Morality, Decision and Social Organization. Toward a Logic of Ethics*. Translated from German by Eric van der Schalie. 1974
   ISBN 90-277-0318-3; Pb 90-277-0319-1

7. Béla Juhos: *Selected Papers on Epistemology and Physics*. Translated from German by Paul Foulkes. Edited and with an Introduction by Gerhard Frey. 1976
   ISBN 90-277-0686-7; Pb 90-277-0687-5

8. Friedrich Waismann: *Philosophical Papers*. Translated from German and Dutch by Hans Kaal, Arnold Burms and Philippe van Parys. Edited by B. McGuinness. With an Introduction by Anthony Quinton. 1977
   ISBN 90-277-0712-X; Pb 90-277-0713-8

9. Felix Kaufmann: *The Infinite in Mathematics. Logico-mathematical Writings*. Translated from German by Paul Foulkes. Edited by B. McGuinness. With an Introduction by Ernest Nagel. 1978   ISBN 90-277-0847-9; Pb 90-277-0848-7

10. Karl Menger: *Selected Papers in Logic and Foundations, Didactics, Economics*. Translated from German. 1979   ISBN 90-277-0320-5; Pb 90-277-0321-3

11. Moritz Schlick: *Philosophical Papers*.
    Vol. I: *1909-1922*. Translated from German by Peter Heath, Henry L. Brose and Albert E. Blumberg. With a Memoir by Herbert Feigl (1938).
    Vol. II: *1925-1936*. Translated from German and French by Peter Heath, Wilfred Sellars, Herbert Feigl and May Brodbeck.
    Edited by Henk L. Mulder and Barbara F.B. van de Velde-Schlick. 1979
       Vol. I: ISBN 90-277-0314-0; Pb 90-277-0315-9
       Vol.II: ISBN 90-277-0941-6; Pb 90-277-0942-4

Vienna Circle Collection

12. Eino Kaila: *Reality and Experience. Four Philosophical Essays.* Translated from German by Ann and Peter Kirschenmann. Edited by R.S. Cohen. With an Introduction by G.H. von Wright. 1979
 ISBN 90-277-0915-7; Pb 90-277-0919-X

13. Hans Hahn: *Empiricism, Logic, and Mathematics. Philosophical Papers.* Translated from German by Hans Kaal. Edited by B. McGuinness. With an Introduction by Karl Menger. 1980   ISBN 90-277-1065-1; Pb 90-277-1066-X

14. Herbert Feigl: *Inquiries and Provocations. Selected Writings, 1929-1974.* Translated from German by Gisela Lincoln and R.S. Cohen. Edited by R.S. Cohen. 1981   ISBN 90-277-1101-1; Pb 90-277-1102-X

15. Victor Kraft: *Foundations for a Scientific Analysis of Value.* Translated from German by Elizabeth Hughes Schneewind. Edited by Henk L. Mulder. With an Introduction by Ernst Topitsch. 1981   ISBN 90-277-1211-5; Pb 90-277-1212-3

16. Otto Neurath: *Philosophical Papers, 1913-1946.* With a Bibliography of Otto Neurath in English. Translated from German and edited by Robert S. Cohen and Marie Neurath, with the Assistance of Carolyn R. Fawcett. 1983
 ISBN 90-277-1483-5

17. Ernst Mach: *Principles of the Theory of Heat. Historically and Critically Elucidated.* English Edition based on the Translation from German by Thomas J. McCormack (1900-1904). Edited by B. McGuinness. With an Introduction by Martin J. Klein. 1986   ISBN 90-277-2206-4

18. Moritz Schlick: *The Problems of Philosophy in Their Interconnection. Winter Semester Lectures, 1933-1934.* Translated from German by Peter Heath. Edited by Henk L. Mulder, A.J. Kox and R. Hegselmann. 1987   ISBN 90-277-2465-2

19. Otto Neurath (ed.): *Unified Science.* The Vienna Circle Monograph Series originally edited by Otto Neurath, now in an English Edition. Translated from German by Hans Kaal. Edited by B. McGuiness. With an Introduction by Rainer Hegselmann. 1987   ISBN 90-277-2484-9

20. Karl Menger: *Reminiscences of the Vienna Circle and the Mathematical Colloquium.* Edited by Louise Golland, Brian McGuinness and Abe Sklar. 1994
 ISBN 0-7923-2711-X; Pb 0-7923-2873-6

21. Friedrich Waismann, Josef Schächter and Moritz Schlick: *Ethics and the Will.* Essays. Translated from German by Hans Kaal. Edited and with an Introduction by B. McGuiness and J. Schulte. 1994   ISBN 0-7923-2674-1

22 Karl Menger: *Reminiscences of the Vienna Circle and the Mathematical Colloquium.* 1994   ISBN 0-7923-2711-X

KLUWER ACADEMIC PUBLISHERS – DORDRECHT / BOSTON / LONDON

BJ 1114 .E76 1994

Waismann, Friedrich.

Ethics and the will